DISCARD

Good Habits, Bad Habits

Good Habits, Bad Habits

The Science of Making Positive Changes That Stick

Wendy Wood

Farrar, Straus and Giroux

New York

Farrar, Straus and Giroux
120 Broadway, New York 10271

Owing to limitations of space, illustration credits can be found on page 305.

Library of Congress Cataloging-in-Publication Data
Names: Wood, Wendy, 1954 June 17– author.
Title: Good habits, bad habits : the science of making positive changes
    that stick / Wendy Wood.
Description: First Edition. | New York : Farrar, Straus and Giroux, [2019] |
    Includes bibliographical references and index.
Identifiers: LCCN 2018060812 | ISBN 9781250159076 (hardcover)
Subjects: LCSH: Habit.
Classification: LCC BF335 .W596 2019 | DDC 152.3/3—dc23
LC record available at https://lccn.loc.gov/2018060812

Designed by Jonathan D. Lippincott

Our books may be purchased in bulk for promotional, educational, or
business use. Please contact your local bookseller or the Macmillan Corporate
and Premium Sales Department at 1-800-221-7945, extension 5442, or
by e-mail at MacmillanSpecialMarkets@macmillan.com.

www.fsgbooks.com
www.twitter.com/fsgbooks • www.facebook.com/fsgbooks

10   9   8   7   6   5   4   3   2   1

This book is not intended to provide medical advice. Readers, especially
those with existing health problems, should consult their physician or
health care professional before adopting certain dietary or exercise choices.

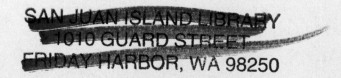

For Steve, who makes everything possible,
even writing a book

Contents

PART I

How We Really Are

Persistence and Change

Habit is, as it were, a second nature.   —Cicero

Every so often, my cousin goes on Facebook to proclaim that she's going to change her life. In her case, that means losing weight. It always starts the same way: she has regrets, she weighs more than she wants, she has a bad back and the extra pounds are making it worse. Then she sums it up in language we can all appreciate. She says she feels *stuck*. She feels like she's unable to change. Lastly, she asks for help from her social media friends.

The world of social media (or at least her small corner of it) is broadly encouraging: "You can do it! If anyone can, it's you."

"There's nothing I know you can't do."

"You're one of the strongest women I know."

"This weight-loss thing will not defeat you."

Her friends bolster her. They successfully play their parts in the sophisticated social process that my cousin is initiating: first, her commitments are shared with her peers, and therefore become stronger and more vivid for her. But there's a second, less obvious step to it: she's also raised the stakes of failing. Her public statements hold her accountable for succeeding. Compared with just a private resolution to lose some weight, her public performance makes disappointment costlier. That's what gives the dramatic edge to these posts. She doesn't just say she would like to lose some weight; she vows that this time she will make it happen. Her friends respond with advice appropriate for a hero starting her journey: "Never believe them when they say you can't." She isn't going to just lose fifteen pounds; she's going to start a new life. Her resolve is clear and strong, and she's made that resolve public.

And yet . . . we all know where this is going.

Classical economics gives us a lens on my cousin's dilemma. *Homo economicus*, or "the economic human," refers to our supposedly immutable and rational self-interest, the kind that would make economic behavior as predictable as algebra. As good exemplars of *Homo economicus*, we are thought to be *utility maximizers*—essentially, we are expected to always be rationally in pursuit of beneficial goals. The notion of this excellent rational figure came into sharp focus about two hundred years ago, in the work of political theorist John Stuart Mill. But even back then, his idea attracted scorn and criticism. In fact, it was early critics of Mill's overweening view of our collective rationality who coined the term *Homo economicus*, to caricature his analysis. Ever since, gradually, and in fits and starts, the field of economics has developed a more realistic and more labyrinthine understanding of human nature. Eventually, even the most fundamental tenets of our economics were amended in light of our stubborn irrationalities. Not even the godfather of modern economics was left alone. It may be true, as Adam Smith said, that we all act in "regard to [our] own interest," but that interest can be defined with spectacular—that is, human—variety.

I couldn't help but think of *Homo economicus* when I saw my cousin's post. If she were a purely rational creature governed by clear intentions, then she could simply and quietly change her lifestyle. No announcement necessary.

How hard is it, really, to change ourselves?

Like most of us, my cousin intuitively knew the answer: it's pretty hard.

So she came up with some proactive ways to commit to that change. She bound herself to her plans and raised the costs of failure. She went beyond simply choosing to change. She started to craft her own social environment into one that made it harder for her to *not* lose weight. This should have worked.[1]

It did. Two weeks after her first post, she updated: down two pounds. "That's a great beginning."

But then: silence.

A month later, she posted that she was still trying, but without much success. "No weight loss to tell you about yet." And that was her last post for a while on the topic.

When I met up with her again six months later, she hadn't lost any additional weight. In fact, the only change was that now she had an additional failure to feel bad about. A costly public one. The end result for her, as for so many people who try to change their behavior, is that it just didn't happen. She had desire, she had determination, and she had some peer support. They're supposed to be enough, but they're not.

The beginning of a solution to this problem is to acknowledge that we aren't fully rational. The reasons behind our actions can be obscure. The things that sustain us can be surprising. Scientists have only recently begun to unravel the multifaceted nature of our selves and to identify our resulting biases and preferences. With this understanding, we can never fully undo these influences, but we can account for them when we act. Our own behavior springs from some of the most mysterious, deeply hidden, and (until recently) unrecognized sources of irrationality.

What's derailing my cousin's attempts to change? What's derailing all of us? The answer is that we don't really understand what drives our behavior. The problem goes even deeper than that. We need to stop overestimating our rational selves and, instead, come to understand that we are made up of deeper parts, too. We can think of these other parts as whole other selves, just waiting to be recognized—and given the command to get to work.

Science is finally starting to reveal why we have been unable to change our own behavior. Better yet, it's showing us how to take this new knowledge and formulate a plan to effect lasting change in our lives.

*

Perhaps you tried to save money by following a budget. Or you attempted to learn a new language through an online class. Maybe your goal was to get out more and meet new people. At the start, your intentions were strong, passionate, resolute. Over time, you couldn't maintain that commitment. And the outcome you wanted just hasn't happened.

This is a common enough human experience: we want to make a change, and we form strong intentions. Supposedly that's all it takes. Just think about how univocal common wisdom is on this subject, from "She just didn't want it enough" to "Are you giving it your best shot?" This facile reasoning begins in early childhood ("Reach for the stars!") and doesn't let up until the very end, that stage of life when many of us will (unfortunately) have to "fight" against diseases such as cancer. The ethos is that your willpower is everything. Self-change therefore becomes a kind of test of our personhood—or at least our conscious part. Nike's famous slogan may have begun with some irony, but the resolute quality of the message—and our receptiveness—has instead made it into the secular commandment that it is today: Just Do It. The corollary is this: if we *aren't* (just doing it, that is), then we must be just *choosing* not to.

I bet that'd be news to my cousin and to all of her friends. She clearly made a choice, and she clearly tried to make it happen. It just didn't. Unfortunately, under these conditions, failure is especially disheartening. Comparison with more successful people becomes painful. It's hard not to contrast our own failures to change with people who are highly successful at persisting in their commitments: professional athletes who train for hours every day; musicians who spend months preparing for a performance; successful writers who continually turn out page after page until they complete a project. We see these super-performers and can interpret their mysterious and enviable success only through the lens of willpower: they must be Just Doing It. But why, then, can't we? Why do our life achievements look puny next to theirs?

We end up feeling small.

It's easy for each of us to conclude that we just don't measure up, that if we just made a strong enough commitment to change, we, too, could be thriving. But we didn't have that willpower. We couldn't Just Do It.

This has become a national phenomenon. When Americans are surveyed about the biggest barrier to weight loss for the obese, lack of willpower is cited most often.[2] Three-quarters of us believe that obesity results from lack of control over eating.

Even obese people themselves report that their own lack of willpower is the biggest obstacle to losing weight. Eighty-one percent said that lack of self-control was their undoing.[3] Not surprisingly, almost all of these respondents in the survey had tried to change. They had dieted and exercised, but to no avail. Some had tried to lose weight more than twenty times! Yet they still believed that they were deficient in willpower.

Three-quarters is a big majority. About three-quarters of Americans presently understand that the earth revolves around the sun. In other words, it's established fact. Willpower deficiency is the problem.

And yet, my cousin's story is hardly unique. I bet that every single one of us has had a similar experience. Every single one of us has failed to evidence willpower. Yet we continue to believe in it. We assign it astronomical authority when it delivers astrological results. What's the missing ingredient that makes real, lasting change possible?

*

This is the puzzle that initially attracted me to the study of behavior change: Why is it easy to make that initial decision to change, and even to start to do some of the right things—but difficult to persist in the longer term? As a graduate student and young professor, I saw some of my most motivated and talented colleagues struggle with this dilemma. They wanted to achieve, and they started interesting projects, but they couldn't meet the challenge of continually being productive in the highly unstructured university environment.

Early in my career, a bright graduate student who had a problem with procrastination joined my lab. He excelled in the classroom but seemed to get lost when working on self-directed research projects. I tried to help him by setting up regular work times and small milestones for completion. He ultimately came up against a hard university deadline. In order to continue, he had to submit his thesis proposal by a given date. On that morning, I showed up early to the office, hoping to read his work, and I was greeted by the picture of a tombstone he had hung on my door. I understood. He never met the deadline and abandoned his dreams of an academic career.

If you have ever spent time in a university setting, you quickly learn that intelligence and motivation have little to do with getting things done on a regular basis. So what does?

It seems to me that the willpower hypothesis comes from an initial error—but in many ways, a rational one. When my cousin decided to lose weight, or when you decide to switch careers, it feels like the most important component has been accomplished. The world is a noisy, chaotic place, inhibiting us from making critical decisions. Most of us avoid making those decisions until we have to. So when we do, it feels like a triumph. We lose a few pounds, we make the job switch . . . but then things slow down. Willpower isn't the issue. If you'd asked my cousin whether or not she still wanted her goal a few weeks after that initial post, I'm sure she'd say yes (although probably with a bit more hesitation).

*

Science is showing that, regardless of Nike ads and conventional wisdom, we are not one single unified whole. In psychological terms, we do not have a single mind. Instead, our minds are composed of multiple separate but interconnected mechanisms that guide behavior.

Some of these mechanisms, it turns out, are suited to handle change. These are the features we know—our decision-making ability and willpower. These are familiar because we consciously experience them. When we make decisions, we consciously attend to relevant information and generate solutions. When we exert willpower, we actively engage mental effort and energy. Decisions and willpower draw on what we call *executive-control* functions in the mind and brain, which are thoughtful cognitive processes, to select and monitor actions. We are mostly aware of these processes. They are our subjective reality, or the sense of agency that we recognize as "me." Much as we experience the stress of exerting physical strength, we are aware of the heavy lift of exerting mental strength.

Executive control must be paid its due. Many of life's challenges require nothing more than this. A decision to ask for a raise at work starts with setting an appointment with your boss. You carefully phrase your request and outline your reasons. Or, you decide to add some romance to your life by asking that attractive person at the gym to meet for coffee. After some deliberation, you find an appropriately casual way to do so. Decisiveness works in these one-off events. We make our decision, steel our resolve, and muster our strength to follow through.

Other parts of our lives, however, are stubbornly resistant to executive control. And thinking every time we act would, in any case, be a highly inefficient way of conducting our lives. I'll return to this later, but can you imagine trying to "make the decision" to go to the gym every single time you went? You'd be condemning yourself to rekindling the ardor of Day One every single day. You'd be forcing your mind to go through that exhausting process of engaging with all the reasons that you felt you should be going to the gym in the first place—and, because our minds are wonderfully, irrationally adversarial, you'd have to run through the reasons *not* to go, too. Each time. Every day. That's how decision-making works. You would constantly be in the throes of heavy mental lifting, with little time to think about anything else.

What we're going to discover in this book is that there are other parts of our mind, parts that are specifically suited to establishing re-peated patterns of behavior. These are our habits—better suited to working automatically than to engaging in the noisy, combative work of the debate chamber that usually accompanies our decision-making. What we'll also see is that a whole lot of life is *already* contained in those automatic parts—the simple, assiduous parts of yourself that you can set to a task. What could be better than that for accomplishing important and long-term goals? Skip the debate chamber and get to work. That's exactly what habits are for.

Science and our own experience have shown that our minds nat-urally form habits, both innocuous and consequential. I bet the first fifteen minutes you're awake goes about exactly the same each morn-ing. That's natural. But it's easy to conclude that our minds must be constantly creating and re-creating active, deliberate tendencies to persist. It's easy to believe that persistence comes from our repeated, conscious efforts to shape our actions to meet our goals. If our pat-terns of behavior were the result of Just Doing It, as too many of us believe, then our conscious minds must be choosing to keep doing the things that it does every day . . . right?

They might if we forced them. But our conscious minds have little contact with all kinds of things we do—especially habitual things. Instead, a vast, semi-hidden nonconscious apparatus is at work, one that we can steer with signals and cues from our conscious mind, but one that ultimately runs on its own, without all that much meddling from executive control. These parts of us are vastly different from the conscious selves we know, and can be utilized in hugely different ways.

The self we know is concerned with raises and romance. Our *non-conscious* selves are forming habits that enable us to easily repeat what we have done in the past. We have little conscious experience of forming a habit or acting out of habit. We do not control our habits in the same way as we do our conscious decisions. This is the under-the-surface, hidden nature of habit. It explains how our casual conversa-tion on the subject is marked by an odd sense of submission: "Ah, well, that's just my habit"—as though habits almost exist separately from us, or run in parallel to the selves that we experience. And it's true,

habits have been a mystery, stuck for decades in the idea that breaking bad habits or forming beneficial new ones is simply about intentions and willpower.

Before we go further, it's important to highlight that the same
learning mechanisms are responsible for our good habits, meaning
ones that are aligned with our goals, and for our bad habits, the ones
that conflict with goals. Good or bad, habits have the same origins.
They result in very different experiences, of course, but don't let that
color how you think of them. In this regard, going to the gym regularly and smoking a couple of cigarettes a day are the same. The exact
same mechanisms are at work.

But for our health goals, exercising and smoking are polar opposites. The purpose of this book is to show how we can use our conscious understanding of our goals to orient our habitual selves. We can
set the agenda; we can direct. If we know how habits work, then we
can create points of contact between them and our goals so that they
sync in astonishingly advantageous ways. They already do in some
cases, as we'll see.

*

I trained as a graduate student in one of the top attitude research labs
in the world. We presented people with information on a specific
topic and tested whether it influenced their judgments and opinions.
We developed powerful models of how people go about changing their
attitudes and behavior. Our focus was on the initial stages of change—
how to influence people to adopt new views of the world. We studied,
for example, the ways that persuasive appeals create support for environmental policies. It was important, valuable work. As I said earlier,
many life decisions are primarily subject to executive control, the
cockpit for initial changes in our lives.

But other things require more than initial decision-making and
will: becoming a better parent, a more responsive spouse, a more productive employee, a more diligent student, or a more prudent spender.
These changes don't happen all at once. Instead, they play out over
long periods of time—years—with actions that have to be constantly

maintained. If your goal is to reduce your environmental footprint, it's not enough to take the bus home from work tonight. You have to do so today, tomorrow, and into the future. To become solvent and pay off your debts, it is not enough to forgo buying those new shoes or that new phone. You have to resist making purchases repeatedly, at least until your accounts are in the black. To form new relationships, you have to persist even if the first person at the gym turns down your coffee invitation. You have to meet more people you might like and repeatedly make offers to connect with them. You have to somehow become committed to the consistent *procedures* of doing things.

I quickly realized, when starting my own research, that persistence was special. I didn't actually set out to study habit; I wanted to understand how people persist. The conventional wisdom was that persistence required strong attitudes—strong enough to get people to make a change and then stick with it in the long term. I realized that it was possible to test this idea on a grand scale by reviewing all of the research that had measured what people wanted and intended to do— sign up for a class, get a flu shot, recycle, take the bus—and then tested what they actually did. Did they follow through on their intentions and sign up, get a shot, recycle, or take the bus? It seemed a simple, obvious question, and one that should have a straightforward answer.

I, along with one of my students, Judy Ouellette, systematically reviewed sixty-four studies including more than five thousand research participants. What we found was surprising. For some behaviors, people acted as expected. If they said that they intended to enroll in a class or get a flu shot, then they typically enrolled or got a vaccine. For these one-off, occasional behaviors, conscious decisions ruled, and people with strong attitudes just did them. The stronger their plans, the more likely they were to perform the action. But other behaviors were puzzling. With actions that could be repeated often, like recycling or taking the bus, intentions didn't matter very much. Thus, people might want to recycle their trash or take the bus to work in the morning, but their behavior did not follow. If they typically threw everything in the landfill, then they continued to do that, regardless of their intentions to recycle. If they usually drove to work, then they carried on doing so, despite their intentions to take the bus. With some behaviors, people's attitudes and plans had little impact on how they acted.

These results were unexpected. It should have been the case that once people made a decision to act and formed a strong intention, they just did it. When I went to publish the results, the journal editor asked me to redo the analyses, but I found the same thing again. So they asked for a whole new study validating the results. Again, we discovered that repeated actions were different. People could consciously report strong attitudes and plans, but they continued their past actions regardless. Finally the research got published, and it has since been replicated hundreds of times. Of course, not all scientists were convinced. Some argued vehemently against the findings, believing that conscious attitudes and intentions are sufficient to explain behavior.[4]

That early research proved pivotal in identifying the special nature of persistence. By special, I mean that persistence wasn't connected to what we had previously thought. It didn't seem to be connected to anything in accepted models, and it didn't follow the formula suggested by conventional wisdom. Persistence seemed to be *more* than what we assumed it was, and somewhat stranger, too. It turned out that we couldn't just conjure it up by asking people to state their intentions. Persistence mostly did not reflect strong attitudes and plans.

But the critics were correct in a way, because my initial research did not explain what *does* lead people to persist. We knew it was special. We didn't know how to make it happen. It has taken decades, but that criticism has finally been addressed. We now know that it's habit that creates persistence. This book explains what we have learned about how to create habits.

*

The myth that behavior change involves little more than strong intentions and the willpower to implement them has been operating successfully for a long time. So it's useful to think about it critically. Exactly how would employing executive control work in implementing long-lasting change?

We know that when people are really decisive and committed to losing weight, it is possible for them to lose fifteen or twenty pounds. This is the amount of weight an obese person can expect to lose over a six-month weight-loss program.[5] That counts for something.

But we know more. Eventually, most people in such programs fall back into their old eating and exercising patterns. Five years after taking part in a typical weight-loss program, only about 15 percent of participants have kept off even ten pounds.[6] The vast majority are back to their original weight, or have even gained more. That counts for nothing.

Commercial weight-control programs are aware of these data. I talked with David Kirchhoff,[7] former president and CEO of Weight Watchers, about the long-term success of their members. He admitted, "In the great majority of cases, when making change efforts, people just can't stick with it. You know, anybody who does Weight Watchers long enough will ultimately be successful—if they're actually doing the program. What we saw was that most people don't. This is the other side of Weight Watchers."

To stay on a program like Weight Watchers involves constant struggle. "I think about it sort of like this," Kirchhoff said. "If you have a weight-loss problem, you will always have a weight-loss problem. If you're wired to overconsume, if you use food in a certain way, if you struggle with food because your metabolism is set a certain way, it is a chronic condition that never goes away. There's no cure for obesity. Which means, periodically, you're going to fall off the wagon. Then you need to get back on track. It's not like you go through Weight Watchers, lose your weight, and it stays off—you're done."

This is a difficult way to go through life. As Kirchhoff reported, "In so many Weight Watchers meetings, you saw the struggle and the pain. You saw people who would lose a hundred pounds. Then they would gain it all back. You saw the impact that it had on them. They feel horrible. They feel like a complete failure. Their confidence is just shaken to the bone."

Weight control is a particularly useful example only because it can easily be quantified and because it has been widely studied. But the same dynamics are at play if you're trying to spend more quality time with your kids, or to save money, or to stay focused at work.

The problem is that the strong-intentions-and-willpower theory of self-change drastically underestimates the likelihood of backsliding. Let's consider how my cousin would try to persist in losing weight through the sheer force of her decisions, without developing new habits.

She will be making that decision in a hostile environment. She regularly buys a lot of junk food for her teenage kids. The result is a kitchen full of crackers, chips, cookies, soda, ice cream. Food is everywhere—sitting on counters, in the cupboards, and packed into the fridge and freezer. In this environment, joined by constantly snacking kids, she eats while watching TV, talking on the phone, and entertaining family. She likes to go to the mall and always makes time for a fast-food break. Her life seems to revolve around eating while doing.

It's worth noting here that the natural environment is not inherently hostile. Our ancient ancestors would surely be amused by the idea that food was anything but scarce, and that one day we would be bedeviled by its superabundance. But the problem isn't just abundance. According to David Kessler, the former FDA commissioner, the food industry is not just aiming to satisfy its customers.[8] The industry, including the growers, concocters, testers, packagers, marketers, distributors, and retailers, is investing in *hyperstimulating* foods—creations with the power to keep us eating. There are scientists hard at work right now devising ways to get you to eat more than you naturally desire. It's important to know this, not to develop a sense of powerlessness, but to preserve our sense of self despite repeated failures. Today's environment poses a big challenge, and we'll meet it and win only if we are able to take its full measure.

Compounding this challenge, my cousin lives in a suburb that does not make it easy to exercise. Her town was built for driving, not walking. She has three cars in the driveway, only a few short steps from her front door. And her house is cozy, without room for bulky exercise equipment.

To follow through on her intentions in this environment, she would have to continually resist the lure of overconsuming and under-moving. Her life would become one difficult decision after another. Every day would feel like Day One, like Groundhog Day: repeatedly resisting the same conveniences and comforts, repeatedly addressing herself to her underlying weakness, repeatedly testing herself.

Decision and will simply aren't the tools to use for making continued sacrifices in order to persist at our new goals. It's too taxing, and would leave us with no time to think about anything else! Even more, the melodrama of this continued self-denial is counterproductive.

Psychologist Daniel Wegner and his colleagues devised an experiment to demonstrate the ironic effect of inhibiting our desires. Participants were instructed in a simple task—*not* thinking of a white bear. Who spends much time thinking of white bears, anyway? Participants sat alone in a lab room for five minutes and rang a bell every time they failed to suppress this thought. On average, they rang the bell about five times, almost once per minute.[9] No surprise that our thoughts wander, even to forbidden topics, when we are alone and bored. What is interesting is what happened when the same participants later sat for five minutes trying *to* think of a white bear. After the suppression task, they rang the bell almost eight times. In contrast, participants instructed to try to think of a white bear for five minutes, but without the initial task of *not* doing so, rang the bell fewer than five times. It was as if the act of trying to suppress a thought gave it a special energy to emerge later. After the participants tried not to think about white bears, thoughts of them returned again and again. When rating their experience, participants who had initially suppressed thoughts of white bears reported feeling preoccupied with them.

This is desire's ironic twist. Trying to suppress it undermines our best intentions and makes our goals harder to achieve. It confounds our good behavior by turning it into torture. As Wegner explained, "We stay awake worrying that we cannot sleep, and we spend all day mentally in the refrigerator when we are hoping to diet."[10] Exerting control has an "oppositional quality that always seems to haunt attempts to direct our minds."

At this point, when our unmet desires loom large and our motivation is at a low ebb, our conscious, thinking selves jump in. Consciousness is facile and easily comes up with justifications for quitting. Excuse making is a talent at which our conscious minds excel. In the moment, you can rationalize eating last night's pizza (you missed lunch) or skipping the gym today (your knees hurt). This talent allows us eventually to stop fighting ourselves and our environment. We are back to where we started.

*

Our lives could be very different if we took advantage of the emerging science on how, when, and why habits work. For something so integral to the human condition, our habits are paradoxically counterintuitive. As you will see, this unknowability is a defining feature of habits that helps make them successful at what they do: persist despite our conscious intentions to do otherwise.

Our conscious, aware self—the part of us we experience moment by moment when we make decisions, express emotions, and exert willpower—is the part we encounter every day. We have the ability to introspect, but we run into the philosophical conundrum of applying our own perceptual and cognitive apparatus to understand itself. We can only know the knowable parts of our experience.

Habits work so smoothly that we hardly ever think about them. The world of habit is so self-contained, it makes sense to think of it as a kind of *second self*—a side of you that lives in the shadow cast by the thinking mind you know so well. Understanding how this works requires the full resources of psychology and neuroscience.

Every once in a while, of course, our habits attract conscious thought. After making a resolution to talk and not message with coworkers, we trash that angry email we automatically started writing. When we remember to conserve water, we turn off the shower. We remind ourselves to put down our phones when having dinner with our kids. We are engaging in executive control, or *top-down processing*, by controlling our unwanted habits with our better intentions.

This is the way many of us live. Our conscious decision-making self is pitted against our habitual, automatic responses. We are wrenched over and over by bad habits, in a sort of internal war.

But there is another way.

We can change unwanted habits and form good ones that are consistent with our goals. When our automatic response is the desired one, our habits and goals are in harmony. We no longer have to rely on will. This is the payoff to this book: understanding how to form good habits amid the pitfalls of daily life. We can learn to form habits that efficiently work with us, not against us.

The truth is that many of your virtues are already habitual. Perhaps you automatically lock the front door when you leave the house? Use the turn signal in your car when you are about to change lanes or make a turn? Kiss your kids every day before they go to school? You might think that you do these things because you intend to do so. More likely, such regularly repeated responses are habits. Habits proceed so efficiently and quietly, we think we must have consciously decided to perform them.

When habits and goals are in line, they smoothly integrate to guide our actions. Most of the time, we aren't even aware that it's happening. We act out of habit without having to make a decision to do so.

As we will see, the habitual mind is in many ways less impressive than our conscious, thinking self. It certainly attracts less attention. But it works with great efficiency. We mindlessly respond to environmental cues, in a kind of *bottom-up processing* of the world as we find it. Walking into your office—check your schedule for the day. Holding an empty bottle—throw it in the trash. Hearing the doorbell ring—open the door. This is the effortless, habitual way of persisting to meet our goals.

What behaviors do you want to change? Maybe you want to have regular family dinners? Establish channels of communication that are more open with your employees at work? Save money for retirement or your kids' college? Indulge more often in cultural offerings available to you? All these can be integrated into that part of your life guided by habitual behavior. They can become what you automatically do. Habits work for us in ways that our conscious decisions never can.

2

The Depths Beneath

The diminutive chains of habit are seldom heavy enough to be felt, till they are too strong to be broken.
                                                                    —Samuel Johnson

But what actually are habits?

Much of my work has addressed exactly this question. Before learning how to promote good habits and break bad ones, we have to understand how they function in our lives.

I became interested in habits after establishing the special nature of persistence. But they are tricky to study because they are inherently unknowable to the person performing them. How could we be sure we'd receive clear, cogent information from participants about something whose utility was wrapped up in its being hidden from our conscious mind?

After a lot of starts and stops, I heard of a research technique called *experience sampling*, in which participants report on what they are doing as they are doing it. It was a novel way to collect data. The in-the-moment quality of this approach suggested that it could capture the experience of acting out of habit, assuming that such a thing really did exist.

We started with a small group of students at Texas A&M University.[1] Each received a stack of pocket-size booklets plus pens to carry with them. They also got a wristwatch that was programmed to beep every hour. At the signal, they were to stop and write down what they were doing and thinking. One student reported, for example, "I am watching game shows, so I am thinking about the answers." Another reported they were attending class while thinking, "I'm really tired." They also rated on a scale how often they had performed that behavior in the past in that context—in the same time and place.

With today's technology, we would just program participants' phones to pose questions. Our crude watch alarms created unique challenges, like what to do with the watches while sleeping. Many participants ended up burying them in a dresser drawer, so as not to be disturbed by the hourly beeping.

After two days, participants turned in the booklets.

Thirty-five percent of the reported behaviors were performed almost daily and in the same location. These actions were routine, but did it make sense to call them "habits"? Could we say that eating, exercising, or working on the computer was habitual? Our premise was that a true habit needed to be performed automatically, without conscious direction. To assess this, we asked participants to record their thoughts while acting. Much of what participants reported was mundane. Someone in the act of "cooking" was thinking, "Did I already add the pepper?" or "I am so hungry." These thoughts corresponded to actions. Participants were monitoring their actions as they performed them or explaining to themselves why they were doing them in the first place. On the other hand, if they noted thinking something like "Hey, *Seinfeld* comes on in thirty minutes" while they were cooking, we confidently coded that action as being performed automatically, without conscious direction.

This combination approach—capturing routine behaviors as well as thoughts concurrent with those behaviors—revealed *how* participants went about performing the behaviors they were repeating routinely. The results were surprising. For a full 60 percent of actions, participants were not thinking about what they were doing. They were daydreaming, ruminating, planning. For example, when exercising, one student wrote down the thought "Where would I like to go for spring break?" Perhaps imagining the sunshine and poolside mojito was a kind of analgesic to the pain of exercise. But the lack of thought about the exercise itself revealed no conscious link there. The mechanics of action were not taking up space in the conscious mind. This is not the repressed, Freudian version of the unconscious, but it is another way that our minds function outside of our awareness.

That's not to suggest that people never thought about their habitual behaviors. Although few of us think much about brushing our teeth, sometimes we definitely do (before going to a big meeting or when we run out of toothpaste). And we found one particularly interesting common trigger for people to become self-conscious of their habits: being with others. Just being around people is enough to turn the spotlight inward and to start to monitor what you would normally do without much scrutiny at all. This is potentially useful if you ever feel like you just aren't very aware of your habitual self (and would like to be): Go public. You'll have a better sense of self in no time.

Back to the study: as you'd expect, the most common habits were showering, brushing teeth, dressing, going to bed, and waking up. Those were the events that took place most often while the participant was thinking about other things. That hardly adjusted scientific understanding. But other findings certainly did. We expected that people would differ when it came to how much of their behavior was ruled by habits. Some might have lots of habits, their days structured around working, eating, socializing, and exercising. Others, we thought, would be free spirits with less structure. This wasn't just born of our own experience; it's a well-established cultural belief and a foundation of classic stories. You find the poles in Jules Verne's Phileas Fogg, whose clockwork daily regimen is structured down to the footstep, and Margaret Mitchell's Scarlett O'Hara, whose improvisational savoir faire keeps her just buoyed on the crest of wave after wave of catastrophe. We expected to find Foggs, O'Haras, and a spectrum in between.

We did not. No personality differences explained how much of participants' lives were guided by habit. Individual character didn't matter. Everyone seemed to rely on habit to about the same degree. It was time to retire that particular expectation.

Another interesting finding was that pretty much everything was subject to habit: 88 percent of daily hygiene, like showering and getting dressed, was done habitually. Fifty-five percent of tasks at work were habitual. Lifting weights, running, playing sports—about 44 percent of those were performed habitually. Resting, relaxing, sitting on the couch—about 48 percent were habits.

Even entertainment could be consumed on autopilot: when participants watched TV repeatedly in the same context, they were likely to be thinking about something other than the show. It seems that we don't have to pay attention to something in order to be entertained. For repeated TV shows and music, only sporadic attention was required. This might sound obvious or familiar, but I realized it hinted at a quality of habit that hadn't yet been well studied: habits are relentless. Television shows are highly formal conglomerations of professional writers, actors, and advertisers, all of whom have done everything they can to capture and keep your attention. Modern television represents the cutting edge of human creative diversion. And yet even this artful enticement will get submerged, eventually, by the force of habit, freeing your conscious mind to think about that meeting on Wednesday afternoon that you're dreading.

For a second study, we asked participants to list not just a single action and thought, but *everything* they were doing and thinking at each beep. They might have reported, for example, talking on the phone while working on the computer and listening to music. With these more complete reports, the estimate of habit was slightly higher, with 43 percent of behaviors performed out of habit.

This was the first research on the daily experience of habit, and we wanted it to be correct. We worried that these findings reflected something about our participants, given that college students' days are tightly organized by class schedules. That structure, we supposed, could artificially create habitual patterns. We decided to conduct the study yet again with people of all ages. We could then see whether people rely on habits more or less across the life cycle. For this final study, we went to a local gym and recruited from fitness classes.[2] We enrolled people from seventeen to seventy-nine years old. Everyone went through the same procedure: booklets, watches beeping every hour, and two days of reporting. We tested for age differences, but we did not find any. We tested for personality differences, but again, personality did not influence habit.

A few new insights did emerge from this additional study. People who had full-time jobs lived slightly more structured days. A greater percentage of their actions were habitual. Working long hours created more repetition in recurring contexts. People who lived with others, especially children, had slightly fewer habits. The influence of others kept people flexible, it seemed. This makes sense. Other people in our life simply amplify the rate of chaos. They get sick, get promoted, go on vacation, make a mess, and generally disrupt our routines. However, when people with all of these different lifestyles were included in the estimate, the total percentage of actions ruled by habit was slightly over 43 percent, which essentially replicated the college student research.

The media, blogs, and popular books reported this research widely. In fact, they homed in on a feature of the work that we hadn't really expected to be the most exciting part: it became widely reported that we had estimated the simple frequency of habit performance in daily life. And that number was extraordinary. Fully 43 percent of the time, our actions are habitual, performed without conscious thought. We had provided the first scientific estimate of how often people act out of habit. It turned out to be a lot higher than science at the time had assumed.

But I was left with a nagging feeling that my work hadn't yet delivered. We had hoped to pull back the curtain of consciousness to reveal the mechanics behind repeated actions. But, really, we had learned more about what habits *aren't* than what they *are*. We traced around the habits in people's lives, and produced a very large section on our map of self-knowledge—but an empty section nevertheless. We now knew that a large part of people's lives is dictated by habit, but we still had no idea how habits actually formed.

Further insight would have to wait. However, I did leave that research project with one important clue to what would come next: we learned that you can make pretty much any behavior more habitual, as long as you do it the same way each time. When we speak casually of habits, we are most likely referring to a specific category of behaviors that we popularly agree are habits, such as brushing teeth, sending a follow-up email, or pulling out our credit card at the cash register. But the category of habits is much broader than we imagine. In fact, it has no real boundaries.

What I began to realize was that habit refers to *how* you perform an action, not *what* the action is. This insight would have consequences.

*

What we *don't* know about habits has already filled books: history books, economic texts, health guides, marriage manuals, and many of the personal diaries sitting in our drawers—all full of our historical, scientific, and personal misunderstandings of why we keep doing the things we do. Online blog posts and bestselling books offer seemingly plausible but mostly scientifically uninformed advice about how to develop effective work habits, healthy eating habits, happy marriage habits, good parenting habits, and prudent financial habits. They rarely note the key feature of habit: it works outside of our conscious awareness.

Only occasionally do we realize that we acted out of habit. Usually, we notice the habits we don't want—overspending (yet again) at the mall, biting our nails, or binge-watching shows late at night when we have to be up early. We also notice others' irritating habits, and we wish that they were more aware of what they are doing. Perhaps a co-worker is routinely late to meetings, eats loudly at his desk, or fails to pick up her trash in common areas. We notice such unwelcome habits in ourselves and others because they get in the way of our current goals. Perhaps reflecting this greater attention to unwanted habits, the Google search engine to date has logged about 291 million searches for "bad habits" but only about 265 million for "good habits." Bad habits get noticed.

But the habits you know, especially the unwelcome ones, aren't the most important habits in your life. The habits that are really driving your behavior go largely unrecognized. Remember 43 percent? If I asked you right now to list all of your habits, would they add up to anywhere near that percentage of your daily behavior? There's no chance.

That's not just because we fail to see some of our buried habits; it's also because our conscious self often takes credit for habits we've noticed and deemed good. We assume that, out of love for our children, we read to them each night before they go to bed. We believe that, out of the desire to save money, we check the specials each time we enter the grocery store. We think that, out of safety concerns, we buckle our seat belts whenever we get in the car.

Psychologists call this overriding confidence in our own thoughts, feelings, and intentions the *introspection illusion*.[3] With this cognitive bias, we overestimate the extent to which our actions depend on our internal states. We are immersed in our own sensations, emotions, and thoughts. These compelling internal experiences drown out our ability to recognize other possible influences on our behavior, especially nonconscious influences such as our own habits. As a result, we are overconfident that we are acting on our intentions and desires. It seems likely that this phenomenon underlies the mystery of our habits. Our curiosity about ourselves has already been satisfied by the belief that we do the things we do because we "will" them. It's flattering and empowering, but it's also false.

The introspection illusion is measurable. In one study, researchers asked people passing through a retail store to identify the best-quality item from four identical pairs of nylon stockings.[4] Given that the stockings were identical, the task should have been impossible. Nonetheless, consumers went through the items, comparing each to the other. In the end, they preferred the rightmost stocking four times more often than the leftmost one, on average. They gave many different reasons for their selections, but no one spontaneously mentioned the position of the stockings. When asked directly, virtually all shoppers denied that they were influenced by item position. According to the researchers, many accompanied their denials with "a worried glance at the interviewer suggesting that they felt either that they had misunderstood the question or were dealing with a madman."[5] The researchers speculated that the choices were influenced by "the consumer's habit of 'shopping around,' holding off on choice of early-seen garments on the left in favor of later-seen garments on the right."[6] Despite showing no awareness of this habit, shoppers still acted on it. In so doing, they were left without a clear explanation of their choices. To the conscious self, it makes sense that we choose based on other things, such as the appearance and texture of each item.

Habits are not the only nonconscious influences we overlook when explaining our behavior. College students, it turns out, even overlook the desire to earn money when it is not at the forefront of consciousness. In an experiment, some students read a description of another student's plans to earn money. In a later part of the study, participants chose between two trivia games, one with the title "American Politics" and the other "American Government." One of the games had pictures of money in the description. After reading the initial story of earning cash, students tended to choose whichever trivia game included the picture of money. It was as if the initial reminder of money was guiding their later game choices. Rationally, this doesn't make sense. No money could be earned, regardless of the game chosen. But, as we saw in Daniel Wegner's white bear study, we can be primed to fixate on almost anything—and money is surely a more seductive concept than bears. Most interesting is that students seemed largely unaware of this influence. After reading the initial story, they did not report a heightened concern about money. Also, when rating a list of possible reasons for their game choice, participants gave least importance on average to their desire to earn money and to the picture of money in one of the game descriptions. They claimed the most important factor was their interest in the game topic, politics vs. government. Once again, the conscious self overreached and discounted nonconscious influences on actions. We make assumptions about what, plausibly and flatteringly, is responsible for our actions.

Our overly generous attributions to our conscious experience make sense in some ways. Many of our habits are useful, and we might have acted similarly if we were thinking carefully about what to do. A sequential-comparison-purchase habit, for example, is efficient. There is no reason to reevaluate items in a display when they are all equally good. It makes sense to just choose the final item we consider. The illusion arises when we fail to recognize the nonconscious habits we are following and instead introspect and unwittingly confabulate explanations for our actions.

There is another way in which over-attribution to conscious intentions can be explained. In doing so, we reconcile ourselves to our choices. They make sense to us. We imagine a better color, texture, or quality in the last item we consider, and we do not question our choice. Or we are attracted by irrelevant features of a task (politics vs. government), and then we are content with our preference.

But there's a huge downside. If our noisy, egotistical consciousness takes all the credit for the actions of our silent habitual self, we'll never learn how to properly exploit this hidden resource. Habits will be a silent partner, full of potential energy but never asked to perform to their fullest. Our conscious self's intrusion is keeping us from taking advantage of our habits.

*

In one of the first studies testing whether voting could be a habit, I worked with the political scientists John Aldrich and Jacob Montgomery to analyze eight national elections between 1958 and 1994.[7] We were looking not at habits of voting for a political party or any given candidate, but simply at the act of going to the polls and casting a vote. People don't vote all that often, so it's not an obvious habit, but even this behavior shows habit-like tendencies.

In a democracy, a lot rests on who casts a ballot. It can literally determine a country's health, wealth, and happiness. Political scientists have developed sophisticated models to explain why some people turn out to vote and others do not. The models follow our intuitions: voters go to the polls when they are highly motivated to do so, perhaps because they are concerned about the election outcome, feel that they can make a difference, identify with a party, or have been contacted by a party. Without these motivations, voters stay home.

The election data revealed whether citizens voted in a particular election, their feelings about the election, and how often they had voted in the past. But only some people, we found, voted when they cared about an election. The political science models (and our intuitions) did not hold for citizens who had voted a few times in the past. These people continued to vote even in elections that they didn't care about. It seemed like they were forming habits that kept them automatically going to the polls. The simple frequency with which people had voted in the past was thus an initial indicator of whether they were acting on habits or conscious decisions. Stronger habits emerged with more frequent voting.

Voting behaviors are useful for studying habits because we vote regularly in controlled ways, and there are robust records of our actions. They're good data. But the hidden operation of habit in voting is intriguing. Voting, in a representative democracy, is one of the three moments when each of us gets counted. The others—the census and taxation—are passive. Things are taken from you (information, money). Voting is different. Your self comes into the picture, and you assert your own preferences and vision for the country. In a democracy, voting is a moment of unification. You and the rest of the nation are briefly connected, during which time you are invited to express your wishes for how the nation will continue to function. Whether we vote, along with whom we vote for, should be the perfect example of *motivated reasoning*, as our decisions are guided by our political values. Accordingly, research shows that thinking about politics engages neural areas involved in emotion and decision-making.[8]

And yet, even in this moment, habits can rule. There simply is no scene where habits cannot enter.

There was another part to the voting study. It may seem obvious at first, but the implications were huge: by moving house, people disrupted repeated, habitual voting. It seemed to make them think more consciously about the act of voting. After a move, regular voters acted in the way most of us intuit and voted only if they were highly motivated. This makes sense, because moving turns voting into a hassle again. When you move, you have to reregister to vote in your new location. You also have to learn new ways to go about voting, such as finding a new polling place or maybe bringing your driver's license. You are no longer automatically repeating what you have done in the past.

Context pervades our understanding of habit. If the *context* remains stable—you keep living in the same place, you keep driving the same route to work, you keep sitting on your couch every evening—then you repeat past actions automatically. These are rich environments for the cultivation and perpetuation of habits.

*

The invisibility of habit hides a huge amount of power over our behavior. Not just huge—also hugely important: the kinds of behavior that habit governs are matters of life and sudden death. Consider how habit benefits us on a weekly trip to the supermarket. You've probably gone hundreds of times. Same car, same road, same destination, maybe even the same grocery list. This environment is a perfect opportunity for habit to take over. In that ten-minute drive, we easily pilot a four-thousand-pound amalgamation of carbon, steel, and plastic, and then put complicated geometry into practice as we ease into that last parking spot. All done on autopilot, with skills learned through repetition.

But sometimes the unexpected happens in the familiar territory between supermarket and home, exactly where our minds might be wandering. Maybe a kid's ball rolls into the street, and she runs after it. Or an elderly couple takes longer than expected to navigate the crosswalk. Or another driver misjudges the light and speeds through the intersection.

A delayed reaction to one of these events can make for tragedy. More than half of all auto accidents occur within five miles of home, during a local trip—going to the grocery store, the laundromat, or any of the countless stops in our own neighborhoods.[9] Of course, we have accidents mostly close to home, because that's also the locale where we do most of our driving. Still, we should be most familiar with the blind corners and challenging intersections in our own neighborhoods—we should be safest there. But in familiar surroundings, habit takes over. We stop paying attention and start ruminating about today's events or planning tomorrow's. Most of the time, we get to the grocery store and back with nothing more eventful than a restocked pantry. Habits make the wildly challenging and difficult seem easy and safe. But driving a car is probably the riskiest thing most of us do on a daily basis.[10]

About 40,000 fatalities occur each year on U.S. roads, along with 4.6 million injuries, whereas driving in Europe is safer, with fewer traffic deaths per capita.[11] The U.S. numbers have been climbing lately, partly due to what is called "distracted driving." We have all been behind the wheel and heard the familiar ping of a text. Do we ignore it? It's tempting to pick up the phone and read it. Rationally, we know the dangers. But driving, especially close to home, feels like second nature. So, many of us pick up our phones, read the message, and maybe even answer it. Five out of ten U.S. drivers in a survey reported reading phone messages behind the wheel, and a third reported writing messages.[12] Even if we withstand the pull of the phone, many other driving distractions capture attention as we select a radio channel, set a destination on our GPS, eat and drink, or reach for an object in the passenger seat.

These are all extraordinarily dumb behaviors. They also showcase the extraordinary potential inherent in habit. It can take one of the most dangerous things we do every day and seamlessly transform it into the background of our lives. Only new drivers, relying on their conscious decisions, feel the adrenaline and rush of fear that all of us rationally should experience on the road. As driving habits form, the wide range of skills required to operate an incredibly complex machine become a background hum behind what we are thinking about—and texting about—each day. Good or bad, habits emerge with practice, and conscious decision-making recedes.

*

So far, we've explored voting habits and driving habits. These are concrete, tangible actions that we can see and understand. It makes some sense that these can be repeated into habits that persist. But what about more elusive, obscure outcomes, like artistic creations? Can they benefit from habit persistence?

An insightful study recruited forty-five professional comedy performers from SketchFest, a large comedy festival.[13] Each was given the setup to a comedic scene and four minutes to generate as many endings as they could. As an example: "Four people are laughing hysterically on stage. Two of them high-five and everyone stops laughing immediately and someone says _____."

The comedians each generated about six funny endings in the four-minute period. (One example: "And that is how the Glue brothers became joined at the palm.") All participants then predicted how many more funny endings they would be able to generate if they had four additional minutes. Their conscious selves expected diminishing returns. The average estimate of new endings was about five, which was fewer than they had produced in the initial four minutes.

They were then given an additional four minutes to work. The actual number of new endings they generated was 20 percent higher than they estimated. They didn't give persistence enough credit.

If they had had a habit of persisting at such creativity tasks, then they would have stuck with the task and produced more ideas than they predicted. Their expectations and desires would not have mattered. With a strong habit of persevering, they would have continued to try to produce ideas, and would have done so successfully, despite their pessimistic predictions.

This same pattern held in other studies involving creative tasks. Like the comedy performers, when college students worked on a task for a few minutes and then estimated their productivity if they continued for a few more, they underestimated the benefits of persistence. They expected decreasing returns for their continued efforts. Amazingly, when specifically instructed to persist, students generated not only more solutions than they anticipated, but also solutions that were more creative. When independent evaluators read the output, the ideas generated at the end of the session were judged to be of higher quality—more creative—than the ones produced initially. Persistence, put to the test, didn't wear down. It just kept on producing. Our misapprehension is understandable. We know that our executive efforts wear down over time. We simply get tired of thoughtfully trying to control our behavior and make decisions. Our attention ebbs and our motivation wanes. But our habitual selves—where persistence sits—are made of totally different stuff. And it's stuff we can put to work.

All of us can make better use of our 43 percent. We can sync up the deep and workmanlike pull of habit with our conscious intentions and long-term goals.

3

Introducing
Your Second Self

Could the young but realize how soon they will become mere walking bundles
of habits, they would give more heed to their conduct while in the plastic state.
We are spinning our own fates, good or evil, and never to be undone.

—William James

A central assumption in my graduate training was that by changing
people's attitudes you can change their behavior. After being convinced
to favor an environmental policy, people should act accordingly by signing ballots and petitions and speaking up in support. This was state-
of-the-science thinking at that time, but I quickly learned it was not
generally shared—at least not by my colleagues at my first job.

Many of my new colleagues were radical behaviorists, and I soon
learned they disagreed with my logic: they called my approach *ex-
planatory fiction*. The first time they said this about my research, I
had no idea what they meant, except that of course to a scientist,
anything that smacks of "fiction" must be bad. It was clearly no com-
pliment. I went back to my new office and read up on the works of
the eminent behaviorist B. F. Skinner. I learned that the fiction, to a
radical behaviorist, was that our attitudes and beliefs work *top down*
to direct our actions. My coworkers rejected the seemingly obvious

truth that concepts in our minds drive our sensations and responses. Their philosophy was very different.

Behaviorism's heyday was in the middle of the previous century. Skinner put pigeons in specially constructed boxes to observe and measure their responses to stimuli. He postulated that humans (and pigeons) learn by responding to stimuli in the environment in order to obtain rewards and avoid punishments. This philosophy quickly became part of the conventional wisdom in the field. To radical behaviorists like Skinner, the idea that our actions are influenced by our attitudes was like saying that we are impelled by ghosts and spirits. A popular metaphor for human actions was a telephone switchboard that coupled incoming sensory signals to outgoing actions. People— through the habits they developed from learning—were supposedly reacting in fixed ways to stimuli around them, driven by rewards and punishments.

But a funny thing happens to conventional wisdom in science. As soon as it starts to take on that mantle, it activates scientific scrutiny. By the 1980s, the field had shifted away from behaviorism toward recognizing that our minds exert top-down control. As a historian of science would note, this shift to acknowledge human agency, or our active, controlling minds, occurred when the children of the 1960s came of professional age, bringing with them their belief in individuals' ability to create social change. In any event, Skinner's star had fallen by the time I began my career. But there were still a few isolated holdouts, including my colleagues at my first job.

In an ironic twist, the initial criticism of behaviorism in psychology had been mounted by a researcher who studied rats in mazes.[1] Edward Tolman, a psychologist at the University of California, observed that when rats entered a maze without a reward, they explored and seemed to learn the layout and form a cognitive map. When a reward was later added to the maze route, they located it quickly. They were apparently flexibly using their earlier-gained spatial knowledge. The suggestion that rats could repurpose old knowledge and act on it in new ways challenged the very heart of behaviorism. Rats did not seem to be responding helplessly to a succession of internal and external stimuli.

It did not take long for psychologists to reason that if rats use information flexibly, people do as well.[2] This insight contributed to what the field grandly calls the *cognitive revolution* of the 1960s. Cognitive psychology experiments started to show that memory was organized and motivated. It wasn't simply responding to bottom-up associations between stimuli, responses, and rewards. There was plenty of meddling from above, too—useful meddling, the kind that our executive apparatus does very well. We discovered that people learn concepts faster and remember them better when they can categorize them into groups. That's prototypical top-down cognition. For instance, the words "chair," "desk," "sofa," and "table" are remembered better than unrelated words like "shoe," "cherry," "wolf," and "engine." Even more challenging to behaviorists, motivation also mattered. When people are hungry, they attend more closely and remember the words "steak" and "cookies" over "paper" and "spacecraft."

It was a sea change for the field of psychology. Flexible, creative thinking entered the professional fray. The whole field shifted from the study of learning and behavior to the study of the mind.

Unfortunately, the cognitive revolution had its own blind spots. Habits were viewed as too simple for the new perspective, which was poised to capture the heights of human reasoning and experience. Cognitive psychologists ridiculed learning theories as "nickel-in-the-slot, stimulus-response conceptions of [humans]."[3] Studies of human agency and decision-making effectively wiped out the earlier work on human habit. We went from envisioning humans as environmentally driven automata to envisioning them as motivations and intellect acting at will on the environments in which they live.

I soon left my first job and joined another department with more contemporary views. But something about my initial confrontation with behaviorism stuck. Psychology's reigning preoccupation with how people think left little room for studying what people actually do. Stubborn behaviorists even made this point initially, arguing that Tolman left his rats "buried in thought." Clearly, cognition alone was no way to navigate a maze. In psychologists' rush to study memory, they seemed to overlook behavior and the environment. My behaviorist colleagues had convinced me that these were too important to neglect. Understanding people seemed to require a synthesis between these two historically separate camps. We needed to find a way to see the whole maze, rather than just our own preferred corner.

The history of psychological thought about habits suggests that we are on the verge of just such a synthesis. The rise and fall of scientific interest in habits is well represented in the graph below, which tracks how often book authors used the term "habit" compared with alternative terms suggesting top-down facets of human agency: "goal" and "evaluation." Google makes it possible to track fashions in whole literatures by searching how often a given word is used in the many books scanned into its database.

The graph begins in 1890, the year William James published his landmark work *The Principles of Psychology,* one of the first texts on the science of psychology. That was a high point in recognizing habits. James was well ahead of his time with regard to insight into the second self, or that side of you that lives shadowed by the thinking mind you know so well. His suppositions are even more extraordinary in the way they set the stage for many of the subsequent developments in experimental psychology. James famously said, "The more of the details of our daily life we can hand over to the effortless custody of automatism, the more our higher powers of mind will be set free for their own proper work."[4] I find little to quibble with, except that

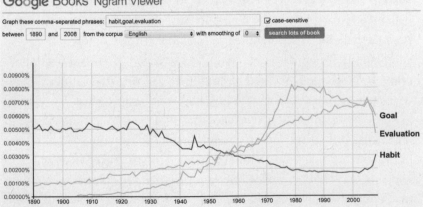

we have a broader understanding of "proper work" than he did, as a nineteenth-century gentleman.

At about the time of the cognitive revolution, habits fell out of favor, at least with book authors. As you can see, authors started to use the word less in the middle of the previous century, as "goal" and "evaluation" gained favor. Psychologists were apparently describing people more as thinking about their aims and purposes than as acting out of habit. The years 1980–2000 were low points for "habit."

The science of habits didn't completely die out, and the rapid upward surge in use of the word during the previous decade is evidence that we are undergoing a correction. What led to this turnaround?

As with so much else in recent years, technology was a driver. Interest in habits resurfaced in part with the development of brain-scan technology (functional magnetic resonance imaging, or fMRI) that enabled previously unimaginable assessments of brain activity. We all can recognize the possibilities in visualizing the performance of the brain, or at least the trace outline of one at work. Other than gazing into a mirror, it's hard to imagine a more literal example of introspection.

The novelty and insight provided by this new technology spurred neuroscientists to study the full capacities of the mind and brain. They started to notice that activity in brain regions shifted as people repeatedly performed a task and started responding more automatically. Technically speaking, when people initially learned a task, their brains showed marked activity in areas involved in decision-making and executive control (prefrontal and hippocampal regions). With repetition, brain activity increased in other neural areas (the putamen in the basal ganglia),[5] as if new areas of the brain got involved with repeated actions. It looked as if we had multiple ways of drawing on neural systems, one to make initial decisions and another to persist.

The habits renaissance was on. At about the same time, research on cognition had started to discover habit-like qualities. One of the most famous studies on attention was actually structured like the habit-learning tasks of behaviorism: see a particular cue, a letter or number, on a screen (stimulus); press a "yes" or "no" button (response); then hear a tone indicating if you are correct (reward). When first learning to do this, participants had to actively make decisions. With enough practice, mental processes became more streamlined. Participants no longer experienced being in active control, they could simultaneously do other things, and they did not have to attend closely to the task.[6] As the researchers explained, participants were acting on "a learned sequence of elements in long-term memory"[7] initiated by consistent stimuli. In this way, habit reemerged in the cognitive revolution under an old Jamesian label, newly rehabilitated: *automaticity*. The new version of habit had a cognitive footprint in long-term memory. It was updated for accuracy and to account for advancements in neuroscience, especially understanding of how the brain works through multiple interconnected networks. Our minds do not just consciously make initial decisions; they also respond repeatedly through habit.

Fundamental insights also bubbled up from research on rats navigating a maze. True, rats are not people, but as we will see, they learn habits in much the same way we do. And early research revealed a central feature of habit: when first learning to press a lever in their cage for food, rats were focused on getting the reward. Researchers concluded that they were goal-directed, with some representation of the reward in mind as they pressed away.[8] If the rewards stopped, then rats did the rational thing and quit pressing. All of this changed with repetition. After much practice with pressing and eating, rats started to act habitually. Even removing the reward didn't stop them. With the lever in view, they just kept on pressing. Researchers concluded that the behavior was driven by familiar cues (lever sight and sound), and the reward had become almost incidental. Of course, after a while even well-trained rats quit pressing levers for no reward. What this revealed is the surprising nature of habits. Habits are a kind of action that is relatively *insensitive* to rewards.

These three streams of research were enough to start turning around accepted wisdom in science. Neuroscientists, cognitive psychologists, and animal-learning researchers converged on a shared recognition of habit, each working independently of the others, and each supplying their unique perspective on the emerging picture.

These developments were fascinating, especially the implication that we can do something once and it's a decision, but if we do it many times in the same way, it becomes something totally different, even recruiting different areas of our brains. It was a model of habit that managed to bring together so much of what we intuitively know: rewards are important when you first do something. We engage executive control and form intentions about what to do in order to snag that reward. Even rats seem to be goal-directed and capable of simple decisions. "I'm hungry, so I'll press this lever and see if I can get fed."

That's just step one. Then, as William James suggested, people act on habit "without any consciously formed purpose, or anticipation of result."[9] Our responses are no longer aimed at seeking outcomes; instead they are triggered automatically by the performance context. This is clearly the case with rats. "I'm at the corner of my cage where I always press the bar, so I'll press," the rat processes, somewhere deep in its brain. But it also works with humans. "I'm standing in my kitchen in front of the refrigerator, so I'll open the door," you process, somewhere deep in your brain. You are no longer consciously deciding that you need something to eat right then: it's habit.

*

Now, finally, it was time to figure out what a habit was. We knew what a habit wasn't—an action that required intention and thought. We knew that repeatedly performing the same task had the effect of reorganizing brain activity. We also knew that habits were prepotent, being primed and at the ready to guide our actions when triggered by familiar contexts. But we were still missing a clear account of what, exactly, was happening in our minds when we acted habitually.

My colleague David Neal and I started with a study of running habits. I was an early-morning jogger, so this project held personal interest. It was a habit developed out of necessity, given that I wanted to have breakfast with my sons before they went off to school and I went off to a full day of work. I had tried exercising later in the day, but with after-school trips to doctor's offices and kids' friends' houses, my exercise always got crowded out. Early morning was the only time on my own. It was a tough habit to start—I remember dreading that 6:00 a.m. alarm the first few weeks. But I loved feeling physically fit, and the regular jog solved my struggle with weight control.

What exactly does it mean to have a running habit? To answer, we recruited Duke University students, some of whom were frequent runners, going out regularly in the same locations, and others who ran only occasionally or not at all.[10] Before showing up for the experiment, everyone had listed a couple of words that represented the locations where they usually ran (if they ever went running). Many noted "forest," given the woods around campus. Some listed "track" and "gym." Participants also gave us words indicating the most important goals motivating them to run (if they ran), such as "relax," "weight," and "fitness."

We wanted to know how people with running habits organized in memory this information about running. So we used a word-recognition procedure from cognitive psychology to test the strength of mental associations between the action (running) and the location (e.g., forest) or the goal (e.g., weight control).

In the lab, a target word was flashed on a computer screen, and participants hit a key on a keyboard as soon as they recognized it. Unbeknownst to participants, a different word was first flashed on the screen prior to each target word. The first word was shown so quickly that participants couldn't consciously recognize it. But their brains caught a brief glimpse. If words are associated in memory, then reading one, however quickly, should bring the other to mind. For example, initially reading the word "coffee" should make it easier to recognize the word "cup." Coffee + cup is a strong, fast mental association. In contrast, reading "comb" first would not speed recognition of "cup."

A running location was flashed as the initial word, and then the second, target word was shown, which was sometimes "running" or "jogging." We measured how long it took participants to recognize these targets. We did the same by flashing a running goal as the initial word and then measuring the time it took to recognize running/jogging.

The results were clear: frequent runners recognized "running" words faster than other participants, suggesting that running was more accessible in their minds. This wasn't surprising. After all, it was a regular part of their lives. But more than general speediness in responding was involved. When their own running locations like "track" or "forest" flashed on the screen first, habitual runners were especially fast to recognize "running" and "jogging." Running was triggered quickly when first cued by *the places they typically ran*. For infrequent runners, in contrast, running locations didn't bring running to mind. Occasional runners didn't have strong mental associations between locations and behavior built up through a history of repeated action in the same context.

Interestingly, flashing habitual runners' goals for running as the initial words didn't speed up their recognition of "running" and "jogging." Weight, say, or relaxing, was the goal that they had said motivated them to work out. But goals didn't seem to be part of the mental associations for strong habit runners. "Weight" or "relax" did not bring running to mind. This fits William James's argument that our reasons for acting become unimportant for habits.[11] It also fits with the findings of my initial research review showing that, for repeated actions, people's intentions and goals don't predict what they end up doing.

Instead, goals seemed important for the *occasional* runners in our study. When one of their running goals was flashed up on the screen, they were especially quick to recognize running-related words that followed. It was as if they had to motivate themselves to run, and so had formed strong mental connections between their goals and working out. For those with a fitness goal, seeing the initial word "fitness" brought running to mind for occasional runners.

Goals and rewards, it seems, are critical for starting to do something repeatedly. They are what lead us to form many beneficial habits in the first place.

We concluded that the speed with which repeated actions are cued by contexts is central to habits. It can keep runners heading out to the track even when they feel tired. If they took the time to deliberate, they might decide not to go that day or to take a shorter route than typical. When people slow down to think, anything might change.

Speed of thought is a clue to how habits gain control. By repeating an action, we change the way that it's represented mentally. We turn an initially motivated action—one that we do to achieve a goal such as physical fitness—into a habit built of strong mental links between performance contexts and our response. When we think of that context, the response snaps rapidly to mind. The payoff of mental speed is that the habitual action is already cued up and ready to go while your slower, conscious mind is still deciding to do something else.

Habit formation works a lot like learning math. When most of us first learn to compute $2+2$, we get the answer by totaling up $1+1+1+1$. But after some homework, we no longer need to do the computations, and instead retrieve the answer directly from memory. That's the feeling of $2+2$ just "looking" like 4. Or of the path by the lake just "looking" like it's time for a jog. When we act on habit, we are essentially retrieving our practiced answers to previously solved problems.

Habit memories are easily put into operation. They simplify our lives by solving the everyday challenges of making decisions in an environment stuffed with choices. In psychology, we call this *chunking*— binding together bits of information into a coherent whole. With a habit of Friday-night takeout at a regular place, you have to remember only one overall sequence, not the multiple separate steps of choosing a restaurant, finding the phone number, placing an order, and navigating there. Or a habit of breakfast with your partner becomes a single unit in memory of the multiple steps: you making coffee, your partner getting out bowls and food, and then both of you making occasional comments about the day as you eat and read media.

Our research thereby linked memory, action, context, and persistence. And a working definition of habit emerged: a *mental association between a context cue and a response that develops as we repeat an action in that context for a reward*. (We will later explore how a mature habit can keep operating without the ongoing presence of rewards.) This definition builds on other well-established mental dynamics, such as chunking and reward learning (more on that later), and adds repetition. That's the analytical and value-neutral definition.

But a shorthand definition is this: *automaticity in lieu of conscious motivation*—automaticity that emerges as we learn from repeated responding. A habit turns the world around you—your context—into a trigger to act.

That easy, fluid, automatic feeling of acting on habit isn't accidental or secondary to the way habits work. Effortlessness is a defining property. The situation you're in triggers the response from memory, and you act. It can essentially bypass your executive mind. The pleasure is in the thing getting done without your consciously lifting a finger. If you've ever managed employees, you should know the feeling well (I hope): you start to ask someone to do something, and she interrupts you to say, "Already did it!"

Put your fingers on a computer keyboard, and you can type without effort. See your child's tears, and you automatically reach for a tissue to dry them.

Although some researchers equate habits with automaticity and assume that they are the same thing, in reality habit is only one form of automaticity. Automaticity has several flavors, just as conscious thought takes many forms. For example, we can consciously form impressions of other people by carefully weighing the pros and cons of being their friend, or we can make a snap judgment, deciding we like them because they are smart or agreeable. Similarly, we have multiple ways of automatically reacting to people, flinching instinctively at their very loud voice (*reflex*) or intuitively liking them for using the same good-smelling soap as an old friend (*Pavlovian conditioning*).[12] Even goals and ideas can be activated automatically by our surroundings.[13] These are all different kinds of automaticity, and each works in its own way. Sometimes these other types of automaticity even become intertwined with our habits (e.g., *Pavlovian-instrumental transfer*). Nevertheless, the habit kind of automaticity is especially important for us because it is a foundation of persistent behaviors.

In our everyday life, we spontaneously learn these different mental associations. Until now, you formed habits naturally as you went about living your life, repeating actions over and over in the same context. Though you are not aware of it, your habitual mind is diligently at work.

It's not picky about what it learns. Just give it repetition, rewards, and contexts.

For example, being the go-to parent at your kids' soccer practice on Saturday morning could make you proud of your parenting skills and community spirit. But it probably represents habits you learned over time. Maybe you came early one day to pick up your kids and had fun talking with the other parents. The coach needed your help picking up equipment. The first few times, you made a decision to help out. She was highly appreciative, and you felt the positive regard from the other parents. After a while, you just seemed to pitch in without thinking a lot about what to do. With enough repetition, you formed a soccer-parent habit while hanging out with friends. Eventually, when you think of the soccer field strewn with equipment, picking up and storing automatically comes to mind. You just do it.

Of course, unwanted habits form the same way. You might be a late-night video game player. Insomnia looms. Your conscious self feels guilty about your lack of self-control. Again, it might just be a habit you unintentionally developed. Perhaps you were bored or restless one night and couldn't sleep, so you started to surf the web and tried out a video game. Do this night after night, and eventually you form a habit of sitting down at your game console instead of going to bed. When it gets late, computer games automatically come to mind. Your habitual self built a problematic habit out of a bit of boredom, a ready computer, and those addictively rewarding video games.

Luckily for us, habits are built on past rewards. In daily life, this is a handy feature. The basic logic of habits is that when we keep doing what we're doing, we'll keep getting what we're getting. Habits are a mental shortcut to obtaining that reward again: just repeat what we did in the past. Rewards can reach through time and continue to operate in the habit formula. This means that we don't have to keep procuring those rewards for ourselves, and it means that even if our values and interests change over time, we don't necessarily need to update the identity of those rewards to keep them current. It's enough that once upon a time, you were rewarded for an action that became a habit.

In psychology, we have a name for the automatic scripts our brains piece together when we repeatedly do the same thing in the same way: *procedural memory*. It's such an important repository of information that only the most frequently repeated patterns get stored like this. It functions somewhat separately from other memory systems, and the specific information encoded isn't accessible to consciousness. This kind of cognitive coding is a sort of mental equivalent of admin-only files on your computer. Your computer's best functioning relies on you not naively messing around in its most fundamental code, which it stashes away behind several layers of obfuscation. This is why we don't know much about our habits. The information we learn as a habit is to some extent separated from other neural regions.

Procedural coding protects information from change. This is the advantage to the way our minds encode habits. You don't forget how to ride a bike regardless of how well you learn to ride a skateboard or surf. You can do it years after stopping. You balance and push the pedals without thinking. While cycling, you can even talk to others or enjoy the scenery. Your bike-riding habit didn't get overwritten by new thoughts and experiences.

Other habits are almost as sticky. Speaking a second language, playing a musical instrument, or cooking a favorite dish are skills that fade only slowly as you fail to use them. Past procedural learning is well preserved.

Other kinds of memories, in contrast, are more vulnerable to change. Episodic memory, or our recollection of specific experiences in our life, is especially at risk. In a courtroom, eyewitness testimony depends on this memory system. It is notoriously unreliable, even when witnesses try to recall the event as accurately as they can. Each time they discuss it with others, they are replacing and changing the original memory trace. It blends with other stories and experiences that the witness is exposed to after the event. For this reason, the most reliable eyewitness testimony is usually the least-contaminated, initial version, especially when witnesses are confident in their early reports.[14]

Habit neural circuitry is very different from this. It is suited to capture recurring responses. Each time we act in the same way, the memory trace incrementally strengthens. Bit by bit, over time, the habit becomes securely stored in procedural memory. Thus, my younger son learned Korean at the military's Defense Language Institute through months of practice. One-time memorized vocabulary words, in contrast, were easily forgotten.

*

Acting on habit has additional benefits. It frees our conscious mind to do the tasks it was designed for, like solving problems. The executive system no longer has to manage life routines. Once we surrender to our habits, our minds are free to perform higher tasks.

Former president Barack Obama and Facebook founder Mark Zuckerberg were well aware of the benefits of relegating routine tasks to habit. They both wore pretty much the same thing every day (except for one fateful khaki day in the West Wing).[15] Obama's presidential standard was a blue or gray suit, and Zuckerberg wears a gray T-shirt. Each identified a wardrobe choice that fit his position, and they simply stuck with it. In a 2012 *Vanity Fair* interview, Obama claimed, "I'm trying to pare down decisions. I don't want to make decisions about what I'm eating or wearing, because I have too many other decisions to make." In 2014 Zuckerberg echoed, "I really want to clear my life to make it so that I have to make as few decisions as possible about anything except how to best serve this community." Now out of office, in a new context, Obama can often be seen in chinos and a checked sport shirt. New role, new clothing habit. Maybe he's embracing the small pleasure of choosing more consciously how he wants to look. He certainly has fewer commitments on his executive mind these days.

These two men understand the duality of our mental capacities. They were taking advantage of acting out of habit—freeing up the conscious mind to tackle the new things that life throws at us. For Obama and Zuckerberg, new challenges involved running the world's most powerful country and the world's biggest social networking site. They could bypass making a conscious decision about what to wear on a given day, while always being appropriately attired for their respective jobs.

Their insights echo those of Alfred North Whitehead, the renowned mathematician and philosopher of the nineteenth and twentieth centuries, when talking about the benefits of math notation, like a plus sign or an equals sign. His 1911 math text explained, "By relieving the brain of all unnecessary work, a good notation sets it free to concentrate on more advanced problems, and in effect increases the mental power."[16] With the right notation, something as philosophically complex as the nature of "plus-ness" becomes a simple, known part of an equation. Our minds benefit in similar ways from good habits. With an exercise habit established or a successful work routine in place, we are freed to make decisions about other opportunities and challenges in life. Habits are notation for our behavioral selves.

*

The upshot of all of this isn't just convenience. The simple cognitive mechanics behind habits, it turns out, save lives in disasters and win games on the football field.

In a classic study, twenty-six fire commanders described how they had tackled an especially difficult fire.[17] The officers were seasoned, with an average of twenty-three years' experience. They recounted a variety of incidents, including fires in residences, hotels, businesses, and an oil-pumping station. When fighting fires, you have a lot of choices. The researchers wanted to understand how firefighters weighed available options and selected the best one. Before entering a building from the front, for example, did they consider other, possibly safer ways of entry? Before directing a stream of water at a particular target, did they identify other targets that could be more effective? Detailed timelines were constructed to identify the decision points during each rescue and salvage operation.

In fact, the interviews revealed that officers *rarely* deliberated. Researchers detected few decision points. As they noted, "In almost none of the cases did a (commander) even report making a decision in terms of comparing two or more options and trying to select one."[18] Even when forced to describe their decisions, the officers did not defend their chosen alternative over others.

Instead, these seasoned officers acted with little thought. They identified a set of cues or aspects of the situation that they had encountered repeatedly in past fires. Standard cues included the building layout; the color, amount, and toxicity of the smoke; the rate of change; and wind speed and direction. These cues triggered immediate thoughts of what actions to take, based on past experiences, and the firefighters just acted. Researchers explained: "Options were selected without any reports of conscious examination, evaluation, or analysis. In most cases, the [cues] triggered an immediate cognition of what had to be done, and the action was taken."[19]

Firefighters appeared to react with single-step, automated memory retrieval. Indeed, they seemed to rely on it. Their minds transformed these high-pressure situations into conglomerations of cues and responses. In life-or-death situations, habits provided a way forward.

Firefighting is similar to playing football in that they are both dangerous professions full of physically strong, talented people. But there are few similarities beyond that. At least, that's what I thought until I talked with Clay Helton, the University of Southern California head football coach, about his goals in training.[20] Helton explained, "The whole deal is to eliminate confusion—making decisions. Confusion creates hesitation, and hesitation gets you beat. It also might get you hurt."

According to Helton, "Any time a young man is confused on a particular play, it's going to slow him down because there's doubt. You want players to say, 'I've been through the scenarios so many times that I take my [conscious] brain out of it. I know exactly what to do based on the experiences and the repetitions that I've had.' I always tell the story about Michael Phelps, the Olympic swimmer," said Helton. "His coach used to fill up his eye goggles with water on the last part of his training session—just in case. When he was swimming a race and he couldn't see, he didn't panic or get confused. Done it over and over in every drill.

"During practice, we create the adversity," the coach explained. "Whether it's rushers, whether it's being hit by a bag, an arm going across you, a defensive man trying to reach out and grab you and your jersey. To be able to say, 'That didn't affect me. Coach already did that 172 billion times.' That eliminates what's going on around a player so he can still focus on the most important thing, which is recognizing what the defense is and where the ball goes. He can say, 'That's what I've been trained to do.'"

The thought processes of firefighters and the players on Helton's team are strikingly similar. Both apparently identify a cue and have learned, through extensive practice, the right response. They are able to decipher these cues despite panic, smoke, or the rush of three-hundred-pound defensive linemen. Seemingly slight and thin, the habit mechanism in reality wields a great deal of strength.

4

What About Knowledge?

Knowing is not enough; we must apply. Willing is not enough; we must do.

—Johann Wolfgang von Goethe

Breakfast is a powerful institution. It seemingly perpetuates itself. Almost all of us submit to its ways. Extensive studies have shown that breakfast is consistently the healthiest meal of the day.[1] It has the most calcium and fiber. There's almost no variation in nutrient content of each of our breakfasts from day to day. What you ate in the morning on Tuesday is likely what you'll eat in the morning on Friday.

Foods at lunch and dinner tend to have more problem nutrients, like sodium and saturated fats. Those meals are also the source of most of the day's calories.

Breakfast, for most of us, is a strong habit. And using the tools from the previous chapter, we can see why: we usually eat breakfast in the same context, perhaps in the kitchen or on the road. Repeated context cues activate the same habits, repeatedly. Adding to this, mornings are typically *not* times for conscious decision-making. We're usually in a rush, so we grab something from the cupboard while trying to get the kids' homework into their backpacks. We just act: pour juice, butter toast. Or, maybe we dash out the door without eating anything and stop by a coffee shop on the way to work.

Breakfast is a powerhouse of a habit. It's all context. To see what happens when we try to habitualize food intake without proper understanding of habits, we need look no further than the number five.

*

How many servings of fruit and vegetables should you eat each day? You probably have a ready answer: five. The number comes from one of the best-known public health campaigns ever launched.

It was initiated in the fields of California in 1988 by a savvy director of the California State Department for Health Services, Ken Kizer. California farmers—who produce about half of all the fruits, nuts, and vegetables grown in the United States—were looking for new markets. They found an eager trade representative in the state health services. And, at the same time, scientific evidence was establishing that our lifestyles contribute to many forms of cancer risk. Call it a happy marriage of commerce and science.

According to Kizer, "Beginning in the mid and late 1970s, the evidence became quite clear about the role of diet in preventing cancer and heart disease and other conditions."[2] A 1981 gold-standard scientific review noted obvious cancer risks from being overweight and using tobacco.[3] Even back then, the science was clear: general diet quality and smoking were critical determinants of the risk of cancer.

At that time, however, there was a lot of opinion and not much hard data about fruit and vegetable consumption. But Kizer was undeterred. He got the National Cancer Institute to partner with California agribusinesses, represented by the Produce for Better Health Foundation, and they jointly created the 5 A Day for Better Health program. As with so much in the past few decades, what started in California spread to the rest of the country—and then the world. Ultimately, the World Health Organization adopted it.

In the words of the National Cancer Institute, the number five was clear, memorable, and actionable. It had stickiness. They also lucked into being fairly prescient: a 2014 review of research noted a small decline in people's mortality with each additional serving of fruits and vegetables they consumed a day—up to around five.[4] Consuming more didn't reduce the risk of mortality further.

Initial optimism about the program was high. News reporters were briefed. Advertisements were created with cute cartoons and catchy jingles. Supermarkets put stickers and signs on approved produce. Schoolchildren went on supermarket tours. A national "5 A Day Week" was designated to get out the word. Recipe booklets were distributed. All the efforts really did the trick. By the measures available, the educational program was a stunning success. In August 1991, right before the effort began, the National Cancer Institute and the produce growers conducted a telephone survey. About 8 percent of Americans were aware that they should eat at least five servings of produce daily.[5] By 1997, the results were strikingly different. Thirty-nine percent of Americans knew that they should eat five servings a day. That's a campaign that any political adviser would be proud of.

But this is not a book about campaigns and policy. This is a book about actually changing lives. So the real question is: What about people's actual behavior? The program's purpose was to get people to consume more fruits and vegetables. Did it?

At the beginning of the campaign, from 1988 to 1994, 11 percent of Americans ate five servings of fruit and vegetables daily.[6] Almost a decade later . . . it was still 11 percent. The change in awareness was real; the change of behavior was nonexistent.

In response, the U.S. government has become even more ambitious. Maybe five servings of fruits and vegetables are not enough. The right number is now eat-as-much-as-we-can. Since 2007, the program has been renamed Fruit & Veggies—More Matters. And now the whole month of September is designated as Fruit & Veggies—More Matters Month.

But Americans are still not biting, so to speak. By 2013, only 13 percent of Americans ate the recommended two servings of fruit daily, and 9 percent ate three servings of vegetables. Other countries have been slightly more successful, with 29 percent of Brits, for example, eating five servings a day.[7] From one perspective, this is perplexing.

Americans are frightened of cancer—it's the number one health fear.[8] And the evidence that fruits and vegetables can help us avoid cancer is strong. Indeed, many people now are convinced about the benefits of eating more fruits and veggies. We all know they're great, we know they'll protect us from our greatest health fear, we know what we're supposed to do . . . but our behavior doesn't change. Sound familiar?

Why can't we make consuming fruits and vegetables as powerful and robust an institution in our lives as, say, breakfast?

Actually, we can. We just need to know how. Almost half of food preparation and consumption is habitual. We all eat out of habit. As we learned in the previous chapter, merely *knowing* something won't get in the way of a mature habit—it's that procedural coding that protects a habit from abstract knowledge and judgment. That 43 percent of ourselves will keep on going, regardless of our fears and our sense of responsibility.

It's easy to see how 43 percent of our eating becomes automatic. Eating has all of the basic components of habit formation: it's frequent, it's often performed in similar contexts, and it (at least initially) is reward-based. It's almost archetypally habit-friendly.

Evidence of the habitual nature of eating comes from a meticulous study[9] that evaluated what more than one thousand people ate at each meal across four weeks. At the end of every day, participants noted what foods they consumed and mailed the report to the researchers. To get the basic details, the researchers analyzed the nutritional composition of each item of food—fat, carbohydrates, fiber, sodium, calcium, calories.

As already discussed, breakfast was the standout, both in terms of nutrition and of consistency. Lunch varied slightly, depending on whether it was at the office cafeteria, a restaurant, or our desk. Dinner was more improvisational. In this study, weekends were different.[10] Participants ate slightly more calories, and the high-calorie foods came earlier in the day (thank brunch for that).

Because of its friendliness to habit formation, our eating is also a popular and useful medium through which to conduct studies on habits more generally. There's one study in particular that goes further in showing how very specific, very concrete context cues can silently rob us of our agency.

Researchers provided participants with all of their food and drinks for twenty-two days.[11] In the first eleven days, some participants got normal-size meals. Others got large portions, 50 percent bigger. All participants were told to consume as much or as little as they desired. Then, everyone got a two-week break before the study started up again. For the final eleven days, diets switched. Participants who had normal-size meals got the bigger ones, and vice versa.

When given the larger portions, participants ate 423 more calories per day than when they were given normal-size meals. You might think that, if they had the normal-size portions first, they'd notice the difference and limit what they ate when portions increased. But participants didn't compensate. They just kept eating the same *percentages* of the food on their plate, regardless of portion size, and that caused them to rack up 4,636 more calories across the eleven days with larger portions than on the days with average portions.

In our real lives, portions of food don't change wildly like this each few weeks. More often than not, we're in charge of serving ourselves, either by cooking or ordering what we want. But the beauty of the study isn't in its reconstruction of our eating habits; it's in how the increase in portion size clearly separated the cues that automatically trigger eating—the *relative* amount on our plates—from the internal cues that we believe guide us, namely how satisfied we feel. By separating habitual cues from conscious awareness, the study showed that we eat in response to available cues: as long as there's food on our plate, we keep going.

What's fascinating is that our judgments of how much we are eating are so often wrong.[12] In a cafeteria study, for example, patrons on several days got the standard serving of pasta and cheese (1,800 calories).[13] They consumed almost all of it (1,700 calories average). On other days, researchers had the cafeteria increase the portion so that the dish was 50 percent larger (2,600 calories), and patrons ate 43 percent more (2,400 calories). When surveyed after the meal, all patrons judged that the amount they ate was comparable to the amount they usually ate for lunch. They also said that the servings were the appropriate size for them. Hardly true, unless they were eighteen-year-old long-distance cyclists.

College students eat a lot of fast food—for some, as often as ten times a week. They average more than four times a week, at least according to a study Mindy Ji and I conducted on fast-food habits.[14] We asked students to rate their *intentions* to buy fast food during the following week. Responses ranged from tepid Yes and No to Absolutely Yes and Definitely No. Then, for the next week, they were to log on to our website every night and report how many times they had purchased fast food that day.

The students who reported strong habits—they bought it frequently, at the same time of day, at the same restaurants, as part of their normal routine—kept to their usual practice even if they didn't *intend* to purchase it that week. Their intentions were no match for their habits. Another way to say this is: we often don't realize what our habits are doing. It's as if they're operating parallel to us, just outside our consciousness. Students were repeatedly buying and eating on autopilot. What about the students who had no fast-food habits? It was this group that was guided by their conscious intentions. If they said they were going to refrain, they did. If they said they'd probably eat fast food during the week, then they did. These students had a plan, and without habits to thwart it, they carried it out. The parts of our lives that haven't been claimed by our habitual self are indeed still receptive to our will—and receptive to new habit formation.

The 5 A Day for Better Health program was a bust at changing behavior. It educated us about healthy foods, but it didn't touch the 43 percent of our eating that is habitual. After being enlightened by the fruit and vegetable campaign, Americans went to the grocery store and chose what they always did—perhaps following a habit of skipping the produce aisle entirely. They continued to snack on candy bars and chips. Their choices were not influenced by knowledge about what those habits were doing to their health.

In the end, the campaign was a testament to the striking disconnect between what we know and what we do. This disconnect has deep origins in the human brain.

*

A new car gets assembled in a state-of-the-art factory. The thousands of pieces and materials—steel, aluminum, fiberglass, leather—come in shapes and molds selected by engineers to fit together into one whole. Assembly is simply the physical reconstruction of a car that already existed, numinously, in the minds of its designers. It is an ingenious and efficient product.

The human brain didn't get assembled by plan, and it is neither ingenious nor efficient. It's a wonderfully crazy collage of parts. It didn't evolve all at once, as a single organ. Instead, it developed in fits and starts over the history of our species. New neural regions and mental functions evolved alongside existing ones. As new areas developed, changed, and were perhaps lost, they modified the capabilities of the human mind. As a result, our brains have billions of neurons comprising multiple interconnected parts, each of which may have evolved at various times. Different *neural networks* are specialized to have slightly different functions.

With those high-tech methods like fMRI scanners that we discussed in chapter 3, researchers can track activation patterns in the brain through changes in blood flow. They can evaluate which neural regions are engaged as we repeatedly perform a task—and develop a habit.

It's worth walking through what this process looks like at a neurological level. Change begins with self-awareness, and there's no more literal way to be self-aware than to meet your own neurobiology.

The habit-forming process often begins with our decisions. We form an intention to do something in order to obtain a desired outcome. The first time you try making a new recipe for dinner or using a new app you just installed, you're making decisions and figuring out what to do in order to get what you want. Which ingredient do I add next? What function do I use now? Add the right ingredients, and you get the reward of a great new dish. Hit the right function on the keyboard, and you've successfully entered data to send a message or record an event. You're learning what to do in order to get the reward you seek.

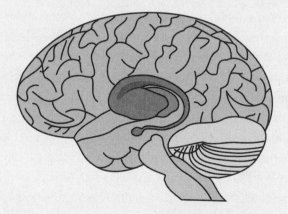

Learning from rewards like this is associated with a neural region called the *basal ganglia,* shown in the picture above. If you were hooked up to an fMRI scanner when first doing these tasks, your brain would show the most activation in a neural system known as the *associative loop.*[15] This involves a part of the basal ganglia, the *caudate nucleus*, along with the *midbrain* and the *prefrontal cortex*, which is associated with self-control, planning, and abstract thought. These executive-control areas are the neural regions your irritatingly consistent and high-achieving coworker uses the most, the one who doesn't seem to need to tap into her second self at all. The rest of us are more occasional users of this choicest of brain regions. Hence, our reliance on some of the other parts to emulate that high achievement.

If you like that dinner recipe and make it again and again, or if the new app becomes a go-to, then your brain functioning shifts. As we repeat actions in routine ways, brain scans show, neural activation increases in the brain's *sensorimotor loop*. This connects a different part of the basal ganglia, the *putamen*, with the *sensorimotor cortices* and parts of the midbrain to form the *sensorimotor network*.[16] Your actions have rewired your brain. You are, to anyone watching you, doing the same thing as when you first learned the action. But your brain is now engaging somewhat different neural systems.

The rewiring makes it easier to repeat what you practiced in the past. You respond more automatically and make fewer conscious decisions. You don't need to check how much salt to add after the flour or remember to hold down a computer key. You no longer worry about how the recipe will turn out or if you'll be able to access a blog post. You have formed a habit.

In a stroke of luck for research, all mammals acquire habits. People, dogs, and whales thrive by learning contingencies between actions and rewards. Our neural systems are structured in similar ways to learn from rewards. With enough practice, all can learn habit associations between contexts and the rewarded response.

Research with rats has yielded many important insights about human habits. And with rats, researchers can use more intrusive interventions than with people. In rats, for example, a particular brain area can be disabled to study effects that we humans would never willingly experience. Many medical breakthroughs that reduce human suffering can be traced to rat studies. Rats have difficulty learning what to do to get a reward after they are given lesions in the *dorsomedial striatum* circuit, an area of the rat brain similar to the *anterior caudate* in humans.[17] Rats disabled in this way do not easily learn to get rewards by pressing a lever in a cage or turning a certain direction in a maze. Disabling other parts of the brain results in very different effects. Rats have difficulty acting out of habit after they are given lesions in the *dorsolateral striatum* circuit, which is a part of the rat brain similar to the human putamen. Despite a history of practice with a maze or lever, once disabled in this way, rats can't use the habits they have learned. This kind of experimentation means that we can develop a sort of atlas of the brain and its common functions, whereas before it was a mysterious, roadless hinterland.

However, our brains are not the same as rat brains. Human brains evolved additional neural regions that allow us to talk, reflect, remember, and plan. We can't compare rat and human abilities to plan, but we can compare some of the ways both species learn habits.

One of the early discoveries in neuroscience that helped to rekindle the field's interest in habit came from a 1990s study that separated habit learning in humans from conscious understanding. The study used the same logic as research that disabled brains of rats, but explored the learning capabilities of patients with specific neural deficits.[18] Twenty participants had Parkinson's disease, which attacks motor control systems in the basal ganglia, especially the putamen, and impedes the ability to learn new habits (even non-motor ones) and to activate old ones. Twelve participants were patients with amnesia who had dysfunction in a different brain area (the *hippocampus*), one that interfered with their ability to remember recent events.

All participants played a game in which they pretended to be weather forecasters. They were repeatedly shown sets of playing cards and were supposed to learn which patterns indicated that it would rain and which predicted that it would be sunny. Parkinson's patients could explain the task and the instructions. They knew consciously what to do. But it didn't matter how much they practiced. They could not learn the connections between cues (cards) and rewarded responses (rain/sun forecast). They could not form a habit.

In contrast, the amnesiacs acquired habits more readily as they practiced the task. After taking fifty chances at predicting the weather, they could make accurate forecasts based on the cards. But when they were asked about what they were doing, they could not remember the instructions or details of what they had seen. It was as if they had little conscious memory of what they had done even though their habits had performed seamlessly.

This research provided some of the first insights into the neural mechanics of habit formation. It suggested that, in humans, habit learning isn't superseded or subordinated by more thoughtful learning systems, as assumed by many researchers during the cognitive revolution. Habits live in resilient, deep-seated neural structures—ones that are fundamental to mammalian life. Our potential for complex, abstract thought helps to separate us from our animal relatives. It's the stuff that we normally think of when we think of ourselves. But it doesn't necessarily follow that what separates us is also most essential to who we are. Studies like these show that our core mental competencies have as much to do with making habits as with making plans.

There are more discoveries to appreciate. Subsequent fMRI scan research has tracked the distinctive neural signature of habit—in the sensorimotor network, especially the putamen—with tasks other than weather prediction, particularly tasks involving sequences of responses.[19] When we learn to tap a keyboard in a pattern over and over again, we learn to connect a cue (e.g., a signal to hit a given key) with a response (e.g., a finger tap). With practice at such tasks, habit neural systems come online with increased activation in the putamen.

Neural studies on habit can get muddled because our goal-directed and habit neural systems are interconnected, and they often work together. We don't have to peer inside the brain to know this is true. Very few describable parts of our lives are purely one thing or the other, and this shading between selves creates ambiguities in empirical findings in habit science. Driving, for example, is a constant sequential trade-off between reacting to the unexpected (conscious thought when another car cuts too close) and habit (context-triggered responding when driving a familiar route). Other tasks likely engage both simultaneously. With your regular Sunday-morning brunch group, the habit circuitry in your brain automatically connects the day and time (context) with stopping by the bagel shop (response) to talk about the no-shows (reward). While headed there, you may be consciously thinking about what you will share with your friends and how they will react. Many of our actions draw on multiple neural circuits like this.

However, ambiguity can be diminished with careful experimentation. Separating habit from more thoughtful action was a chief challenge in my early lab research. Outside of the lab, I could show again and again that people acted habitually, repeating their past behavior instead of doing what they intended. But I couldn't demonstrate habits under controlled conditions in the lab. In my experiments, participants repeated tasks over and over, but when they were later tested, they persisted with the task by making a conscious decision to do so, not by relying on habit. I tried changing the tasks by making them simpler and easier. I tried more training sessions. All ended in failure.

It was frustrating work until I realized the permeability between habit and conscious decisions—and the power of context. I was, in fact, successfully training habits, but during testing, participants were thinking carefully about what I wanted them to do. If you've ever been in a laboratory study, you know the feeling of wondering about the purpose of the research and what it indicates about you as a person. Everyone becomes a practicing scientist in the lab. It was like this during my tests: participants were actively thinking about what they were supposed to do in the study, and they consciously overrode their habits.

My research finally succeeded when I realized that I needed a testing situation that provided a more realistic, lifelike context. Outside of the lab, our motivation and ability to think carefully like a scientist is drained by the many distractions of work, social media, negative people, the news, traffic, bills, and our families, to mention a few. So I added simulated distractions during testing, such as videos to capture and hold participants' attention. I also gave them cognitively draining tasks to do beforehand, to sap their energy and give them something to ruminate about. With conscious decisions so engaged, participants in my research started to act on habit. As in daily life, they just did what came to mind easily. They didn't think so hard about what they should do to impress me.

With habits starting to surface in the lab, we got more ambitious. Guy Itzchakov, Liad Uziel, and I convinced participants in one study that sugar was bad for them—this was an easy case to make.[20] We then gave these sugar-haters a choice of soft drinks, juices, or water. When the lab imitated real life (with a cognitively draining task), then participants ignored their newly healthy attitudes and drank whatever was their habit: If their standard choice was sugared soda, then that's what they chose. If they typically drank water, then they chose that. In a single study, we captured the conundrum that baffled the National Cancer Institute and the Centers for Disease Control in their healthy-diet campaigns. In daily life, it is just easier to act on habit than to make a decision based on our best intentions. But in our lab study, some participants were not subjected to the draining task, and they chose beverages thoughtfully in line with their new anti-sugar attitudes—under these circumstances, participants tended to spurn the sugared soda.

That's a key aspect of why habits are crucial for long-term behavioral change. Brainpower is overwhelmingly costly.

We all know that conscious control is supremely powerful—after all, it's responsible for such major advances in civilization as indoor plumbing and the computer microchip. Why can't it just take over and control our habits as well? The reality is that exerting control is inherently draining, making us feel tired, stressed, and overwhelmed. Control also presents an opportunity cost. We can react to only a few things simultaneously, and when controlling one thing, we necessarily overlook others that could be important. Habits, by virtue of their location deep in the rudimentary machinery of our minds, are relatively cheap. They hum along on barely any bandwidth at all.

Alfred North Whitehead used a military analogy to explain the inherent limits of control: "Operations of thought are like cavalry charges in a battle—they are strictly limited in number, they require fresh horses, and must only be made at decisive moments."[21] Psychologists have a more descriptive name for this occasional use of cognitive control: the *default-interventionist* system.[22] The idea is that we're using the default mode of autopilot most of the time, unless there's a good enough reason to intervene with conscious thought—like, say, imminent flanking by the enemy. Then? Call in the cavalry, by all means. But don't waste their flash on eating enough vegetables.

We exert control over our actions when it's relatively easy and/or when the outcome is sufficiently important to us. This cost-benefit analysis determines whether it's worth it for us to do something other than act automatically.[23] Given the costly nature of control, we use it sparingly.

*

Name the animals in the picture. Couldn't be easier, right? You just look at the animal and say the name. It's hard to go wrong with such simple pictures. To identify the animal, you probably think you look at the image. But the written name is there, too, if you need help. It's a challenge only for a child.

The task becomes more difficult in the next example. Try it—you're probably slower now. You realize that you were doing more than just looking at the pictures. Your habitual response, learned through a lifetime of practice, is to interpret the words you read. This habit—which you probably didn't realize was an influence—now interferes with correctly labeling the animal. Essentially, you have two possible responses, and the wrong one—the habitual one—comes to mind faster. In order to give the correct answer, you have to stop your first impulse and think before you speak.

You've localized and experienced the feeling of acting on habit. When words and pictures align (as in the first task), habit and decision integrate so smoothly, you don't even recognize the habit. But when they conflict (as in the second task), then you have to exert executive control over your habit self.

This is a riff on the classic Stroop task (a test of interference in processing), and it's a simple analogue to the more complicated conflicts we experience between our habits and our current goals. Much like when we try to change our behavior, habits in the Stroop test produce the wrong answer. You find yourself staring at something with a beak and feathers and saying, confidently, "Cat." It's a little harrowing. Identifying shapes and animals is kindergarten stuff, right? Except when there's a simple conflict between two pieces of information: the drawing and the label. Our attempts to respond despite the inconsistency activate brain regions involved in the allocation and execution of cognitive control, especially the *dorsal anterior cingulate cortex*.[24] This part of our brain rapidly registers a conflict (that thing has a corkscrew tail . . . but it clearly says "DOG"), along with the costs (effort) and potential gains in resolving it. Given the ease of the Stroop example, it was probably worth it to you to put in the small amount of additional effort to ignore the label. Voilà, you have the correct answer.

The low-stakes Stroop test is a terrific way to isolate and experience a small example of something that is unfortunately quite a bit larger in real life. Few of us are in situations where we need to rapidly identify farm animals. But in more realistic situations, we do see our desired responses start to erode: when the gains diminish . . . when focus becomes harder and harder at work . . . when you fall behind on assignments, when you slip from three runs a week to two . . . and the benefits associated with your continued effort seem to be getting dangerously low. Will you try to prop those benefits up? By running harder? By trying harder?

Or, instead, will you lower the costs of trying in the first place, by relying on habit?

5

What About Self-Control?

What gets us into trouble is not what we don't know. It's what we know for sure that just ain't so.   —(mis)attributed to Mark Twain

One of the most famous, most publicized, and least well-understood studies in the history of psychology started with a test of four-year-olds' self-control. Preschool children in the Bing School at Stanford University were individually given one small marshmallow on a plate. A child who was able to hold off for fifteen minutes before eating it would get two marshmallows. With these instructions, the experimenter left the child in a room alone.

Children don't have much to do when alone with a marshmallow. Almost 75 percent succumbed to temptation and ate the single marshmallow. The study was conducted again with different temptations, including pretzel sticks and animal crackers, with comparable results.[1] On average, children waited about nine minutes. That meant that most of them lost out on the better prize.[2] All of this is pretty much what you'd expect.

But that successful quarter is interesting. How did they manage to restrain themselves? While they waited, these children figured out distracting strategies. They sang songs. Some fiddled in their chair, in that way only four-year-olds can. Others, when asked, reported that they imagined the marshmallow was a cloud, or a pillow, or anything that they could not eat. Even young children were capable of self-control—if they used the right strategies.

The research gained further acclaim by following the lives of the children across adolescence and into adulthood, revealing that self-control is an enduring characteristic. Those who were able to hold out longer against temptation at age four got better grades in school as adolescents and also scored higher on the SAT. As adults, these kids even weighed less, with a lower body mass index (BMI).[3] *Delay of gratification*, as it was called, seemed to be a fundamental social-cognitive skill, linked inversely to general impulsivity and directly to conscientiousness and executive control—a skill that could provide a lifetime of benefits.

These results were widely reported in the media. Ambitious parents anxiously ran the marshmallow experiment on their own children to predict their future success. It seemed that there was a surefire way to test whether or not a kid would succeed in life.

The research was so iconic that it was incorporated into one of America's best-loved TV shows, *Sesame Street*. Self-control of eating had become socially important given the rise in childhood obesity. Cookie Monster, the voracious blue puppet and eater of all things, especially chocolate chip cookies, was trained to control his impulses. On the show, Cookie Monster played the Waiting Game, in which he got one cookie now or two if he waited. In one episode, he first distracted himself by singing. His song quickly became about how much he wanted cookies. He then imagined that the cookie was a picture in a frame, but found that impossible. He tried to focus on a toy but got bored. He imagined that the cookie was a smelly fish. Finally, enough torturous time had passed, and he won the two cookies.

Cookie Monster's struggles are the essence of self-control. We think of it as resisting temptation, inhibiting impulse, and white-knuckling it through. Cookie Monster's battles with desire were amusing to watch and designed to build character, but he was clearly not having fun.

The children in the original study shared his distress at waiting. If you go online, you can find videos replicating the marshmallow experiment. In one, a cute girl with an orange hair bow tries hard to resist temptation. She handles the marshmallow, smells it longingly, but puts it down. After a short while, the temptation is overwhelming, and she eats a tiny piece, and then another. She screws up her face and looks away, trying to focus elsewhere. Ultimately, there's no marshmallow left. She did not get the reward of a second marshmallow.

There is a critical piece to the self-control story, however, that was mostly overlooked by the media and by *Sesame Street* writers. It revealed a very different side to self-control—a side that is helpful for anyone wanting to acquire a new habit and not give in to temptation.

The Stanford studies actually showed the importance of *situations*. In the original experiments, some of the kids were allowed to see the tempting marshmallow while they waited, whereas others were not.[4] The scenario was the same for everyone—they could still get the marshmallow now if they wanted. But for some, it was simply out of sight. Children were able to wait about ten minutes when the treat was hidden. When it was in plain view, they lasted only six minutes. Those four minutes reveal a lot about the limited power of the self in self-control. Perhaps it's not so much an inherent disposition but instead a reflection of the situations we are in.

The later-life results backed up this power inherent in situations. Being able to wait longer when the treat was hidden did not lead to more successful life outcomes. Waiting was something that was achievable by many. Only when the marshmallow was available, visible, and tempting did waiting signal resiliently high performance throughout life.

The lesson here is optimistic for the 75 percent of us who, as children, could not withstand temptation, and who have continued to succumb as adults. If we are in the right situation, we can achieve similar results to those who are more disciplined. Even if we don't have "it" at a young age, we can arrange our world in a way that enables our success.

There was another option to fight temptations, highly touted by the original researchers. That is, rely on conscious control. Just make yourself think about something other than the tempting thing—those expensive shoes or that electronic gadget that has you hankering. However, cognitive control, as we saw in the previous chapter, is effortful and transitory. Thinking a happy thought might be robust enough to help a kid wait another couple of minutes in a lab room, but it's not clear that it works for very long in daily life.

To study everyday self-control, we asked college students to report every time they felt "Oops, I shouldn't do this"—which happened most often when they slept too much and stayed up late, ate too much, or procrastinated and were lazy.[5] They reported an average of two to three such thoughts each day, and they also reported what they did, if anything, to exert control. When they later reported their success at restraining themselves from temptations, distraction was one of the least effective strategies. The clear winner? *Stimulus control.* Students effectively quashed the temptation when they removed themselves from the situation or removed the opportunity to do that tempting thing. They left their apartment with its comfortable bed in order to study in the library or threw out that last piece of chocolate cake so they couldn't eat it tomorrow. Even for adults, successful self-control came from essentially covering up the marshmallow.

Few of us think that self-control comes from the situations we are in. This is the Protestant ethic at the root of American culture. Puritans believed that self-indulgence was the route to eternal damnation. Through personal denial and deprivation, they were signaling that they were among the chosen few who would get into heaven. It's hard to take this too seriously; Puritans also believed in burning witches at the stake. But the value they placed on personal self-denial has influence even today.

*

The world is a lot more complicated than a test room, and its temptations are a lot more sophisticated than marshmallows. We need to look at what passes for self-control in the real world to understand what it is and how people use it successfully.

First, let's measure ourselves. June Tangney, Roy Baumeister, and Angie Boone devised a self-report scale to measure how much self-control each of us has.[6] This often-used scale supposedly tests our "ability to override or change inner responses, as well as to interrupt undesired behavioral tendencies and refrain from acting on them."

The questions fall into two general groups. One assesses self-discipline (or its absence): "I am good at resisting temptation" and "I refuse things that are bad for me"; or, "I am lazy" and "I blurt out whatever is on my mind." Another set of questions is about the ability to achieve important practical goals in any way you can: "I eat healthy foods," "I keep everything neat," and "I am always on time," or their opposites. (You can see how you score by going to www.goodhabitsbadhabits.org. Check out "How Much Self-Control Do I Have?")

Thousands of studies have used this scale. The results show, much like the marshmallow test *with temptations exposed*, that people who score higher in self-control enjoy more success in life than those who rank lower.

At university, college students who scored higher got better grades.[7]

In relationships, partners who ranked higher were less likely to pick fights.[8] Even the perfect partner might be late for a date, forget a commitment, or overlook your needs. People who score higher on the scale don't escalate these moments into conflicts. They forgive.

Parents with more executive control provide more consistent support and care for their children. When kids act up, as they inevitably do—they're defiant, ignore your good advice, or are just sullen and hostile—parents can react in many ways. Those with greater control were able to modulate their responses and not escalate the drama.[9] They were able to help their kids cope with feelings and learn from frustrating situations.

People who score higher have better credit ratings and save more for retirement, as shown in a study of Swedes. They pay off credit card debt and keep track of their expenses.[10]

People who rank higher also are healthier and weigh less. In one study that followed people in Switzerland over a four-year period, those with greater self-control ate more nutritious foods, overate less often, and maintained a healthy weight.[11]

*

Clearly, it's beneficial to score higher on the self-control scale. You're able to meet a whole range of life goals. But that's as far as the scale goes. It doesn't reveal much about what people do to achieve all these wonderful outcomes. At face value, the items in the scale seem to be measuring the ability to deny oneself life pleasures and be conscientious. This has been the assumption of most researchers: people high in self-control apply white-knuckle tactics to forgo immediate pleasures in favor of long-term rewards—more of that Protestant work ethic.

The first inkling that this might *not* be how self-control actually works came from a 2012 study of Germans from the city of Würzburg.[12] First, they completed the self-control scale. Then they were given personal data assistants to carry with them that beeped seven times each day. Just as in my own study from chapter 2, when they heard the beep, participants responded. In this study, they recorded whether they had experienced any desires or wants in the past thirty minutes. Half the time they were beeped, participants reported having had a desire during the previous half hour. The most commonly reported desires were to eat, sleep, and drink, followed by media use, leisure, social contact, and hygiene-related activities.

Participants also indicated whether the desire (if they had one) conflicted with a personal goal. For instance, a desire to go back to sleep would presumably interfere with the goal of getting to work on time. The wish to eat dessert would be in conflict with the goal of losing weight. About half of the desires were at odds with a goal.

Finally, participants recorded whether they were trying to actively resist those desires. Did they feel like they had to control themselves? Full and half measures counted: Eating part but not all of a chocolate bar. Trying not to mention politics with someone. Deciding to forgo a purchase. Anything that you could look at and say: I did this instead of that. Participants were in general quite successful at exerting control:

when they had an unwanted desire and exerted self-control to stop themselves from acting on it, they were successful 83 percent of the time.

Then the researchers did something ingenious. They matched these results with the results of the self-control quiz that all the participants took before the experiment. Intuitively, what should be the result? We'd expect that participants who scored high on the self-control scale would be a big part of the 83 percent. Self-control is white-knuckle denial, right? And we believe that, starting at a young age, some of us are simply stronger than others. . . .

That's not what the team found. Instead, participants who scored highest in self-control *seldom reported resisting desires,* period. They just didn't experience many unwanted desires in the first place. They didn't have many urges that conflicted with their goals. It looked as if they were able to avoid temptations altogether. They were living their lives in a way that hid the marshmallow almost all the time.

Participants who scored low in self-control had the most battles. They experienced a lot of inconvenient desires that conflicted with their goals. They had to work hard to get their impulses under control. They had to keep fighting temptation over and over again, in a constant, unhappy tug-of-war with their unwanted cravings. Of course, people with really low, bottom-of-the-scale self-control probably just act on all desires willy-nilly and don't try to curb any. Low-control people in this study, however, at least tried to exert control, even if they weren't especially effective.

Effortful self-denial, it seems, is the recourse of people who score low on the self-control scale. They get themselves into difficult situations that require immediate action. But controlling impulses is like the proverbial finger in the dike. It is a short-term solution that works in the moment. These are the people who, in general, do not achieve all of those long-term goals that matter to us—good grades in school, happier partners in relationships, support and care for their children, good credit ratings and adequate retirement savings, and good health and weight.

So, this study revealed that people high in self-control are not living a life full of self-denial and deprivation. Somehow they are managing their lives better. What *are* they doing to achieve all of these wonderful successes in life?

*

I won't bury the lede. They have good habits.

The evidence comes from studies in which people rated their self-control and then reported about a variety of health behaviors: exercise, healthy snacking, and sleep times.[13] As you'd expect, people who scored higher on the scale exercised more, ate healthier snacks, and had more regular sleep and wake times. People who scored lower did not get to the gym as often, ate a lot of unhealthy snacks, and had irregular sleep patterns. These are the standard self-control findings.

Most revealing is how these participants performed these healthy actions. Everyone recalled the most recent instance in which they had, for example, exercised, and reported how they did it. Participants who scored high on self-control said that they automatically went out to exercise without thinking much about it. They usually did it in the same times and places. It had become part of their routine. Once again, people high in self-control were achieving success without exerting much effort. They weren't white-knuckling their way to being healthy.

If you talk to someone who runs six miles regularly, they will tell you that the first mile can be tough. They might mention that the last mile can be tough, too. But once they get started, they don't think much about stopping or whether they are uncomfortable. Someone with a strong running habit isn't thinking much at all about what they are doing. They have a set pattern, and they follow it. They are not making decisions. Here's the very happy implication: the worst, most effortful run will be that first one. Or the second, perhaps. But effort doesn't last (in fact, if it does, you're doing it wrong). Habits will form and take the effort off your hands.

This was also true in the study for healthy eating and sleeping.[14] People high in self-control might take an apple into work every day to eat as their mid-morning snack. Or have a few almonds after dinner in the evening. They routinized eating healthy snacks so that they ate them at the same time and place every day. They did not have to think about it, they just did it.

To get enough sleep, people high in self-control again formed beneficial habits that they acted on without a struggle. They turned off screens in order to go to bed at the same time each night and set the alarm for the same time each morning. They reported doing this automatically, without thinking. They were not struggling with themselves to play just one more round of a computer game or keep reading their Twitter feed. For them, sleep was not a battle of self-control.

Even teenagers exert "self-control" in this way. For 109 teenagers enrolled in a five-day meditation retreat, the day started at 6:30 a.m. and ended around 10:30 p.m.[15] The program had extended periods of sitting and walking meditation, along with small-group mindfulness exercises. Students did not talk for half the day. Cell phones were prohibited.

The actual retreat was just a preliminary thing. Researchers wanted to test whether students continued to meditate once the retreat was over. At the outset, all kids had completed the "self-control" scale. At the end, they indicated on a survey whether they planned to meditate over the next three months—they might intend to do it most days or only rarely. Three months later, you can anticipate the results: kids who scored highest on the "self-control" scale were the ones who met their meditation goals. They stuck with their plans. If they intended to meditate, they did so. As so many other studies have shown, high scorers on the "self-control" scale achieve many positive life outcomes.

But again, the scale did not reflect "self-control" as we commonly understand it. Successful students didn't actively try to inhibit responses to temptations in their lives. They reported that meditation had simply become automatic for them. It was something they did without thinking. They had formed meditation habits. High "self-controllers" achieved desired outcomes by streamlining, not struggling.

*

Research on daily life tells us a lot about how people function. But it's a messy picture, and it's hard to be certain that habit alone creates success. So people's self-control has been tested under lab conditions, where everyone has the same task and is judged on the same scale.

A review of 102 self-control studies assessed performance at various tasks.[16] Some involved beneficial behaviors, such as doing homework, using condoms, and quitting smoking. Others involved more harmful activities, including snacking, cheating, and marital fighting. As you'd expect, people with more "self-control" performed more beneficial and fewer harmful actions.

Beyond this standard effect, the authors of the review anticipated that high "self-controllers" would especially shine at difficult tasks that required the *central executive*. After all, this is what, until recently, we thought the "self-control" scale measures—sheer force of will. But even in these more controlled studies, the data did not support this idea.

Instead, high "self-control" people performed better at the more habitual, automatic tasks than low "self-control" ones. High "self-controllers" were simply proficient at automating. The researchers (who, interestingly, included Roy Baumeister, one of the originators of the self-control scale) concluded that "self-control may in general operate more by forming and breaking habits. It is thus mainly by establishing and maintaining stable patterns of behavior rather than by performing single acts of self-denial that self-control may be most effective."[17]

This cements the shift in our understanding of self-control. People who score high in "self-control" seem to be doing nothing that the scale was ever designed to assess. They do not experience many unwanted desires, almost as if they had neutralized the temptations in their environments. They also know how to form habits by repeating the same things at the same times and in the same places. We'll see how isolated behavioral repetition alone isn't the best way to supercharge habit formation. Also, the overall consistency of the experience is key. Our morning routine in the shower—shampoo hair, shave, soap up, scrub down, dry off—turns out to be the framework for reliably meeting important goals in life. Is there something else we'd rather be doing? We don't even consider the possibility. We follow our shower routine without thinking about alternatives. We do it without struggle and stress.

It's hard to give up the idea that people with good "self-control" achieve so much because of their willpower and self-denial. If you listen carefully to successful people, however, you can start to see the habits that enable them to reliably meet goals without much struggle.

One of the most financially successful people in the world, Bill Gates, founded Microsoft and now has a net worth of around $100 billion. He talks about having to form the right habits to succeed at school and work. In interviews, he admits that he used to be a procrastinator.[18] As a student at Harvard, "I liked to show people that I didn't do any work, and that I didn't go to classes and I didn't care. People thought that was funny," he said. "That was my positioning: the guy who did nothing until the last minute." Gates would cram for tests during the short "reading period" right before exams.[19] This habit worked for him as a student. He flashily demonstrated his brilliance, conjuring mostly A grades from the illusion of working.

But when he dropped out of school after two years, Gates found that the business world wasn't impressed with his shows of instant brilliance. "Nobody praised me because I would do things at the last minute." He admitted, "That was a really bad habit and it took me a couple years to get over that."[20] In the business world, he realized, he had to become more like the students he knew in college "who were always organized and had things done on time." To describe the business habits he developed, Bill Gates uses a flying metaphor: "Pilots like to say that good landings are the result of good approaches."[21] In a similar way, "good meetings are the result of good preparation." He learned to email documents and data to people before a meeting so that they had already analyzed the information. Meetings were more productive, and Gates's colleagues benefited. They were less likely to bear the brunt of his famous impatience.

Self-control is simple when you understand that it involves putting yourself in the right situations to develop the right habits.

＊

Back in my lab, we had moved beyond assessing people's ratings of the strength of their habits. We created new habits.

Along with Pei-Ying Lin and John Monterosso, I wondered if we could build habits that would fill in for self-control—that is, habits to keep people acting in beneficial ways despite temptations and short-term desires.

For college students, especially in Southern California, junk food is a recurring temptation. Sadly, many college-age women base their self-worth on their body image, believing that gaining twenty-five pounds is one of the worst things that can happen to them.[22] The women we chose for this study all wanted to be thin and healthy.[23] Their dilemma was that they really liked M&M's. The question was, could these women learn to automatically choose vegetables over M&M's?

The study used a computer game in which participants tried to win actual carrots (that they got to eat) by moving a joystick toward an image of carrots on the screen. When carrots were available, they always saw the same swirling picture cue, colored purple on the screen, that's the middle image on the following page. Participants played the game when they were hungry, not having eaten for three hours beforehand. The game thus created a very basic, rudimentary habit of moving the joystick in the direction of the carrots when the swirly purple picture appeared. Participants successfully pushed the joystick in the direction of the carrots (in this case, downward), and won (and ate) about eighteen baby carrots each.

Participants returned the next day, hungry again. They played the same game and won more carrots. By this point, they had acquired a habit of moving the joystick toward the carrot picture. Participants did it quickly, without deliberating.

Then came the twist. The game changed. They could win—and get to eat—either M&M's or carrots, depending on which direction they moved the joystick. What would they do? Would they keep making the healthy carrot choice, moving the joystick downward, or give in to temptation?

Participants with strong habits of choosing carrots essentially responded before they had a chance to consider the alternative. The habit (move the joystick downward when the swirly purple picture was on the screen) took control, and participants were on to the next trial in the game. Even when they could get M&M's, participants chose carrots 55 percent of the time. I shouldn't have to tell you that that's far higher than what would happen out in the "real world." People definitely do not prefer carrots to M&M's.

Habits did something special, and with such simplicity. They had essentially replaced self-control. Out of habit, participants ate more carrots than candy. We had set up the experiment to mimic real life—participants completed a draining self-control task before they chose food. With executive control on the wane, participants fell back on their habits.

Habits are not always at the ready to protect us like this. The experiment changed a little more. Now the carrot image was moved to a different spot on the computer screen, as you can see on the following page. And the center picture changed to a spiky brown image. To

choose carrots, participants had to point the joystick in a new direction (to the left, as shown in the image). These small changes shouldn't,
logically, have made much difference. But the easy habitual response
was no longer activated. Participants now had to think about what
food they wanted and which way to point the joystick. Carrots were
no longer the favored choice. They chose M&M's 63 percent of the
time. Even small changes in the cue and response forced participants
to make conscious decisions and to rely on executive control and
willpower.

The study turned popular wisdom on its head. We expect deliberation and willpower to be the route to health, happiness, and success. Indulging in forbidden treats (like M&M's and marshmallows)
should be the action that requires little thought. Instead, when you
have the right habits, the opposite is true. It's when you stop to think
that you might stray from your plans and goals.

If you know how to form a habit, then beneficial actions can
become your default choices. Your best self, your habit, is uppermost
when you are not thinking.

*

The good effects that we popularly ascribe to "self-control" are, it
seems, more accurately captured by *situational* control.[24] The studies
and stories just cited established this mechanism, a mechanism that
will undergird every part of habit formation. A habit happens when a
context cue is sufficiently associated with a rewarded response to
become automatic, to fade into that hardworking, quiet second self.

That's it. Cue and response. Notice that there's no room in that mechanism for, well, you. You're not a part of it, not as you probably think of yourself. You—your goals, your will, your wishes—don't have any part to play in habits. Goals can orient you to build a habit, but your desires don't make habits work. Actually, your habit self would benefit if "you" just got out of the way.

PART II

The Three Bases of Habit Formation

6

Context

Habit is a compromise effected between the individual and his [or her] environment.
                                                          —Samuel Beckett

If you could step out of a time machine into a major corporate office in the 1950s—American Can, Republic Steel, International Paper— you'd already expect some things; we've all seen *Mad Men*. There are few women and no computers. No empty venti cups (but probably plenty of mugs). There's more physical clutter and more paper, but also a lot more room. Open floor plans haven't yet caught on. But there's one thing that shocks you, even though you knew to expect it, even though you were fully, intellectually aware that it would be the case: people are smoking a lot—indoors. They smoke when they get there in the morning, during meetings, during lunch, and on the way home. The (few) women do, too—for them it seems to symbolize a kind of sexual equality (that seems conspicuously lacking in other, more important respects). The men smoke like chimneys. It's supposed to look chic, or manly, or both. Of course they're all hopelessly addicted. You step back into your time machine and return to the twenty-first century. We haven't figured everything out, certainly, but the air quality sure is better.

The 1950s were the heyday of cigarettes in industrialized nations. Nearly half the U.S. population smoked regularly,[1] along with almost 80 percent of the U.K. Many doctors would tell you that smoking in moderation was perfectly fine. Then medical research started to discover what we now know all too well. British researchers, Richard Doll and Richard Peto, provided some of the first evidence that smoking was linked to cancer. Smoking reduces life expectancy by as much as ten years.[2]

In 1952, *Reader's Digest* published an article to that effect titled "Cancer by the Carton." The 1950s were also the heyday of *Reader's Digest*, so many millions of Americans were exposed to that article. The warning was scary, but smoking rates dropped only slightly. Nevertheless, tobacco companies fought back. They attempted to calm any incipient fears by putting filters on cigarettes and by increasing advertising. People kept smoking.

The turning point in the United States came with the famous 1964 Surgeon General's report. The data were clear: tobacco was (and, unfortunately, continues to be) the leading cause of preventable death in the United States. This time was different—kind of. People were finally ready to believe it. Opinions shifted quickly after the Surgeon General's report. Five years after it was published, about 70 percent of Americans recognized that smoking was bad for health.[3] Warning labels were put on cigarette packs in 1966.

But, just as with eating fruits and vegetables, knowledge did not quickly translate into action. Forty percent of Americans were cigarette smokers in 1964. Forty percent were smokers in 1973.[4]

Addiction played a significant role. Nicotine's power to create dependency is often compared with that of heroin and cocaine. But you know how this story ends—not the way it might have, given what happened with the fruit and vegetable campaign and the addictiveness of nicotine. Instead, many people quit, and many more never even started. In fact, only about 15 percent of Americans and 28 percent of Europeans now smoke.[5] Huge swaths of the United States are largely tobacco-free. As a country, the United States cut smoking prevalence by more than half in about fifty years.

The success rate is not as good as it could be, especially among people with low incomes, which is partly due to the large number of tobacco retailers in low-income neighborhoods and the artificially low prices (i.e., discount coupons, offers) in these areas.[6] But the reduction is still impressive to laypeople and social scientists both. It shows that society-wide change is possible. And it shows *how* we can make that change happen.

Informing smokers of the risks had only a mild impact on smoking rates. Even after the Surgeon General's landmark 1964 report of the dangers, U.S. tobacco sales continued to climb until 1980.[7] In taming habits, knowledge is just not a powerful lever.

Willpower also isn't much help—not when stacked up against nicotine. The Centers for Disease Control report that 68 percent of smokers say they want to quit completely.[8] However, each individual attempt usually fails.[9] Only about one in ten actually stop smoking for good.[10] Most end up relapsing, typically within a week. To quit successfully can take thirty or more attempts.[11] Repeatedly trying to quit and finally succeeding requires almost superhuman self-control. To be clear, the fact that some smokers will attempt to quit thirty times or more should not be thought of as a spectacular, ongoing failure on their part; it should be understood rather as a sign of a remarkable underlying persistence. Those people have impressive willpower to continue trying.

You'll already be familiar with what I'm about to say: those super-persistent people simply are not like the rest of us. So, what was successful—for *us*? If knowledge and willpower weren't the answer, what worked? How did so many normal Americans stop smoking?

In 1970, people all around the world were glued to their television sets by the generationally iconic Apollo 13 events. First horror, and then amazement, and then relief: it was something we would never see the likes of again. There was something else Americans would never see on TV again: an advertisement in December of that year with the message "You've come a long way, baby." To imply that smoking was as emancipatory for women as the right to vote, *anti*-suffragettes in nineteenth-century clothes voiced their opposition to both, to the tune of a Gilbert and Sullivan operetta. The ad was selling Virginia Slims, and it was the last time—ever—that cigarettes would be hawked on U.S. TV. We can thank President Nixon, and his signing of the Public Health Cigarette Smoking Act,[12] for that.

Other public manifestations of our nicotine habit also went away. Do you remember cigarette vending machines? Smoking on beaches? In trains? In offices?

Tobacco control laws changed the environment for Americans who smoke. In many ways, those laws literally made the smoking environment smaller and less contiguous. Now smokers have to take the elevator down to the ground floor and file outside. Once the environment changed, so did the habit. We can empirically test this. The fact that each state in the United States has different anti-smoking laws means that we have a set of variables to compare and contrast. This provides a sort of natural experiment to identify the policies that work.

For instance, smoking in workplaces, restaurants, and bars is prohibited in at least twenty-eight states, along with many cities and counties. As a result, about 60 percent of the U.S. population cannot smoke in most places outside their homes and cars, even if they want to.[13]

These bans seem to be effective.[14] Of the ten states with the lowest smoking rates, nine have laws against it in workplaces, restaurants, and bars.[15] The three states with the highest rates of smoking (Kentucky, West Virginia, Mississippi) do not have such laws. In those states, nearly one out of every three residents smokes.

Bans don't change desires. Instead, they put smoking habits in direct conflict with legal sanctions, a conflict that habits tend not to win. A study of sixty-five smokers recruited from pubs in the U.K. is especially good at revealing the nature of this clash.[16] Participants knew that, after smoking was banned, they would be fined if they smoked. But their usual cue to light up—entering the pub and having a drink—continued to be activated. Almost half of those tested in the study unintentionally started to smoke in the pub. For them, smoking was automatic: "Enter pub—light up."

Their comments revealed a struggle: "Yes, once lit a ciggie, remembered, and walked outside." "Yes, I did it last week. It's something I've been doing for years, and old habits die hard." "Put cigarette to mouth but remembered in time. This has happened several times."

The struggle wasn't so much with nicotine. We know that because it didn't matter whether the participants normally smoked a lot or only occasionally. Heavier smokers didn't do any worse than lighter smokers. The culprit was habit and habit alone. At the beginning of the study, before the ban took place, pub-goers were asked if they automatically lit up without thinking about it. These were the habitual smokers. After the ban, strong habit smokers found themselves lighting up inadvertently. Their habit was oblivious to the new law.

A prohibition like the tobacco ban disrupts the automatic "perceive cue; perform response" mechanism of a habit. People who were cued to smoke at the office or in restaurants now had legal reason to curb the automatic response. They had sufficient motivation to consciously override the action of smoking triggered by the environment.

The conflict between habit (smoke here) and conscious awareness (it's now illegal) should decrease over time. As people repeatedly comply with a prohibition, their habits become linked to new places, ones where they now repeatedly smoke. In this case, the pub-smoking habit becomes healthily more inconvenient. Smokers have to stop their conversation, put down their drink, get up, walk outside, and spend a few minutes in some British weather.

Yet another deterrent on smoking habits is cigarette taxes. On average, about half the cost of a pack of cigarettes in the United States now goes to federal, state, and local taxes.[17] Residents smoke less in states with higher taxes. In 2018, Missouri had the lowest tax, a seventeen-cent surcharge;[18] 22 percent of that state's residents smoke.[19] New York had the highest, with a $4.35 charge. Only 14 percent of New York State residents smoke.

For each 10 percent increase in taxes levied on a pack of cigarettes, adult smoking drops on average 4 percent.[20] No special magic there: the more expensive cigarettes get, the fewer we can purchase.

The effects of the environment on cigarette use really become evident with the additional forces that we've legislated. We've already seen that the tobacco companies are barred from advertising on television. It goes well beyond that. In most places, stores cannot advertise cigarettes or put them where customers can help themselves. Buyers have to ask a clerk for cigarettes displayed behind a counter.

We've all stood in line waiting while someone fumblingly describes to a store clerk: "I'll have a pack of Camel Blues . . . no, not those, the 99s . . . no, not those, right above them, the 99 Lights." Doing this every time they smoke becomes another obstacle.

But are all these changes really enough to cut consumption of something as addictive as cigarettes? It's easy to get hooked on nicotine. Is it really possible that a few inconveniences can fight back against that?

Exposure to tobacco cues was evaluated in a study with 475 actively quitting smokers in Washington, D.C.[21] For the month of the study, participants reported each day how much they craved a cigarette. As you'd expect, many relapsed and started smoking again on days when their cravings were high. Cravings fill our conscious minds and direct our decisions.

But that's not what was novel about the study. The aspiring quitters agreed to have their locations tracked on their cell phones. The D.C. area is geocoded, so researchers could tell when participants were close to neighborhood stores selling tobacco. Participants went to these stores for many reasons, including to get gas, to buy groceries, or even to purchase a pack.

If you are like most people, your mental model of a relapsing smoker is: They pick up a cigarette after a long bout with their cravings. The urge builds . . . and they lose the battle. The researchers themselves predicted that relapse happens when you add cravings to the opportunity to purchase. Just flip the polarities, and you have a model for my cousin, the aspiring exerciser: her will to jog diminishes, and it loses the battle with her desire to just relax. However compelling, these models do not capture the ways our actions are maintained out of habit. They better describe how we respond to momentary temptations.

For smoking, relapse actually worked like this: quitters could enter a local store reporting zero craving. That is, they had checked 0, or "Not at all," to the question "Right now, how much do you want to smoke?" When the store sold cigarettes, quitters were exposed to familiar purchase cues. Maybe they saw another person buying a pack. Maybe they glimpsed their chosen brand in its usual place behind the counter. *These cues alone led to relapse*, and the quitters left the store with a pack in hand. They were back to smoking.

The implications for tobacco health policy are clear: we should celebrate laws that restrict point-of-sale cues. No more cigarette machines offering packs as you enter a restaurant. No ads flashing on a screen with cigarettes displayed. No others in a bar lighting up. Despite the addictive power of nicotine, cues in our everyday environment play a big role in whether or not we smoke. Performance contexts make smoking easy or difficult in ways that our conscious self does not understand. Disrupt the scenery of smoking and you disrupt smoking. If we want to counterattack against the ravages of cigarette use, we shouldn't charge head-on into its most advantageous weapon, addiction; we should flank it and cut it off.

Tobacco control was a stunning success. It has many lessons for us.

*

The famous early psychologist Kurt Lewin believed that our behavior is influenced by forces, much like objects in the physical world are subject to gravity and other fundamental forces.[22]

Some of the pressures that act on us come from inside ourselves, in the form of our goals, feelings, and attitudes. This is the part of our worlds, or *life spaces*, that reflect us, as people. If you want to start getting more sleep, for example, then that desire is a force that *drives* you to go to bed early and remove screens from your bedroom. If you decide that you need to work late one night, then that would be a *restraining* force on your sleeping, one that keeps you up.

For Lewin, the *contexts* we are in (which he called "environments") also generate forces on our behavior.

Context refers to everything in the world surrounding you— everything but you. It includes the location you are in, the people you are with, the time of day, and the actions you just performed. Even your mobile phone represents a context that is a physical as well as a virtual space external to you. These are the external forces that drive or restrain our actions. Thus, in Lewin's famous equation, behavior is a function of the person and the context/environment. To get technical, we would notate it like this: $B = f(P,E)$.

Restraining forces are like a kind of *friction* that impedes action. Friction plays a big part in our material lives—when we hit the brakes while driving, strike a match, or walk down the street, we're relying on friction. It also features in economic thought. Economists lament the friction that arises from the time, effort, and costs between a provider and a customer, slowing transactions and producing inefficiencies.

Lewin used these *force field* principles to explain when we will change our behavior. In his terms, tobacco control laws are *restraining forces* that increase the friction on smoking. But other aspects of our context can *drive* smoking, too, by reducing friction. You might see other people light up, which reminds you that you haven't had a cigarette in a while. Whether an external force is driving or restraining, imposing friction or removing it, depends on the behavior and forces in question.

We can think of our lives in terms of their own force fields. Yes, each of us is a source of some of these forces, but the contexts around us are also powerful in driving or restraining actions. We deliberately take advantage of some friction-reducers in our lives. We know it's easier to save money with a regular automatic transfer from our paycheck deposit to our savings account. Despite the initial pain, we eventually don't really notice the reduction in take-home pay. By automating driving forces, we repeatedly save at each pay period.

Marketing appeals are some of the most obvious forces in our daily contexts. A classic driving force, designed to reduce friction on purchases, is "Would you like fries with that?" This simple request at the end of every takeout order encourages us to eat more fried food. We may find ourselves saying yes even if we didn't intend to buy anything.

Driving forces are also responsible for much binge-watching on Netflix or Hulu, as the next show starts without you having to move a muscle or make a decision. The media context just drives you forward into the next episode.

Retailers continually create new forces on us to buy, like integrating purchases across digital and physical stores. As a shopper, you can immediately buy an item you see online and pick it up at a local store. The driving forces link the convenience of online shopping with the immediacy of acquisition, along with the bonus of saving on shipping cost. Retailers benefit from your initial buying impulse online as well as from your potential purchase of additional items when in their store. There's even a name for browsing and buying across multiple channels: *omnichannel* retail.

Ride-sharing companies, like Uber and Lyft, were designed on the low-friction principle. As Professor M. Keith Chen, former head of economic research for Uber, explained to me,[23] it was supposed to "be a one-button-press product. When you open the app, the phone's GPS knows where you are . . . you don't even need to think about it. Just hit a button and say, 'I need a ride.' The car arrives, you get in the car, you tell the driver where you need to go, you get out without handling cash. That was the initial app. You never even saw a price."

He went on to explain: "Everybody called it 'frictionless,' it was a popular term in Silicon Valley. You just want to make this as close to magic as is possible. To early users, it was magic. I can hit a button on my phone and suddenly someone pulls up and they just take me wherever I want to go? That's amazing."

But surge pricing changed all that. "We had the wrong psychological frame around it from the passenger's perspective," said Chen. "It was like a penalty. It broke that frictionless blueprint. 'Aw, man, red lightning bolt appears, and I've got to think to myself, 1.6 times normal, what's going on?'" So Uber changed their pricing practice. "Now riders just see a price. We don't even show you the ridiculous surge. Now it's just, 'Hey, you want to get from A to B, that's going to cost $11.64.'"

Lewin's insight about contexts as force fields has more power than he ever imagined. He recognized driving and restraining influences that we can use to our advantage.

There's perhaps no simpler context influence we can engineer in our lives than sheer proximity. Proximity determines the external forces to which we are exposed. We engage with what is near us and tend to overlook what is farther away.

Controlled lab experiments highlight the importance of proximity to what we eat. Imagine that you arrive at a lab kitchen to participate in a taste test. The interviewer escorts you in and then leaves the room, saying, "I will be right back with some questionnaires. By the way, that food is for you, if you'd like something to eat." There are two bowls. One is full of buttered popcorn. The other holds apple slices. You are left alone for six minutes.

One day when you stop by for the study, the bowl of popcorn is on the table within easy reach, about a foot away, while the apple slices are on a counter—visible, although you'd have to stand up to take one. On another day you stop by, the apples are on the table and the popcorn is on the counter.

So what do you do? You can eat anything, and it makes sense that you eat more of the one you really want (probably popcorn), regardless of location. But this is another case in which our intuitions aren't accurate.[24] How much more popcorn would you eat if you didn't have to get up to grab it? In the study, a lot more. Participants ate about 50 calories when the apples were within easy reach, but about three times more when the popcorn bowl was within reach. Friction in this study was pretty simple—distance. Just putting the high-calorie snack slightly out of reach was substantial friction. Participants could still see and smell the popcorn, but the distance was enough to discourage eating.

I saw this kind of friction in action during a scientific conference on habits I run each summer. One year we had a lot of Europeans attending. I requested extra fruit because they seem to like it more than Americans do. The food service provided extra fruit, but they positioned it in a box to one side, where people had to reach to get a piece. Once I realized this, I moved the fruit to the end of the line, where it could be easily grabbed. It all went immediately, despite the bananas' by then well-ripened state.

This same kind of distance friction is powerful when we are purchasing food in a cafeteria or buffet. In studies that vary where foods are placed, diners favor the more visible, easily accessible items.[25] By putting desserts at the end of the line (instead of at the beginning) and making healthy foods easier to see, restaurants can influence what people eat.

Grocery stores recognize this external pressure. We are its pawns every time we shop. As the saying goes, "Eye level is buy level." If we have to bend low or reach high, we're less likely to bother. We have all become used to grocery stores with end-of-aisle promotions, staples like meat and milk at the back of the store so that you have to walk through the aisles (as you check out those eye-level products), and candy and magazine temptations at the checkout counter where you wait. Can you even imagine a store in which milk and meat were right up front, the cheapest items were at eye level, and apples were by the checkout counter? That store's chief concern would not be profit—by exploiting you and your worst impulses. It would be your good health and well-being—by serving you and your best ideals.

There would be good reason for such a store. City residents tend to eat more fruit and vegetables when they live closer to supermarkets.[26] This holds especially for groceries that devote more shelf space to produce.[27] Farm stands are a good example.[28] In the summer of 2010, the Sustainable Food Center in Austin, Texas, built temporary farm stands in low-income neighborhoods that did not have easy access to fresh fruit and vegetables. The researchers didn't try to educate residents about health or even advertise the farm stands. Instead, they just observed the effects of the increased proximity.

Several weeks before the study began, the researchers screened neighborhoods within half a mile of the intended farm stands. About 5 percent of residents reported ever buying from a farm stand. On average, residents ate around 3.5 servings of fruit and vegetables a day. The stands were then installed outside schools and outside community centers where food stamps were distributed (the farm stands accepted stamps).

Two months later, almost a quarter of the residents initially contacted had purchased from a farm stand. More important, actual fruit consumption doubled, and residents consumed slightly more green salad, other vegetables, and tomatoes (or salsa fresca—this was Austin, after all). On average, residents in the survey increased their fresh produce consumption by about 10 percent, to over four servings. The stores around us can do much, it seems, to promote our health.

Can something as simple as proximity get people to exercise? Between February and March 2017, a data analytics company looked at this question using cell phone records from 7.5 million devices (yes, our phone use is being evaluated in ways we are only starting to realize). They analyzed how far people with mobile devices traveled to paid fitness centers.[29] People who covered a median distance of 3.7 miles to the gym went five or more times a month. Those who traveled around 5.1 miles went to the gym only monthly. That seemingly small difference—less than a mile and a half—separated those who had an exercise habit and those who went rarely. To our conscious minds, such a small distance does not make sense as a barrier. But it clearly was associated with whether people exercised habitually.

Distance may even determine who your friends are. A classic 1950 study evaluated the friendships that developed among 260 married veterans in a student housing project at MIT.[30] People were randomly assigned to live in apartments in small two-story housing blocks at the beginning of the school year. Researchers measured the distance between everyone's front doors. Then they tracked who became friends with whom.

Students did not randomly link up and form friendships. They were much more likely to become friends with their next-door neighbors and with people who lived on the same hallway than with those on another floor. Residents of units separated by as little as 180 feet never became friends. Those living in end-of-corridor units were also less popular, because they didn't meet up with as many people in passing. And the only students who made friends with people on other floors were the ones who lived close to stairwells.

If you think about it, this study suggests a way that we can use external forces to social advantage. If you are moving to a new town and want to get to know people, you can enlist driving and restraining forces to help. Finding an apartment close to a shared building entrance will naturally put you in contact with others. At a new job, choosing a desk that is centrally located, maybe by the office coffee station, will reduce friction on meeting people. And if you have children, they are natural friction reducers, connecting you to the neighborhood through their school activities. You can think of these forces as kinds of "rip currents" that carry you into desired experiences and away from undesired ones.[31]

*

Social forces are created by other people in our contexts. What they do or don't do influences our own behavior. We consume more around heavy eaters than light eaters, regardless of whether they are actually present or we just learn about their eating.[32] But we don't always recognize this influence. Even when quite evidently influenced by others' choices, research participants typically report that they were driven by hunger or taste of the food, and not by others' behavior.[33]

The friction created by others extends to military academy cadets' almost fanatical commitment to fitness. About thirty-five hundred U.S. Air Force Academy cadets were randomly assigned to live together in groups, providing a natural experiment in which they did not choose their roommates.[34] Lazier cadets could not decide to live with others so inclined. Just as a result of the randomness in this assignment, some units had cadets with higher fitness scores in high school and others had cadets with lower scores—although, of course, everyone was reasonably fit.

Students spent the majority of time interacting with the thirty others in their unit. They lived in adjacent dorm rooms, ate together, and studied together. Across the first two years, students with lower fitness scores when they entered the Academy actually reduced the fitness of others in their unit. That is, cadets were more likely to fail their own semiannual fitness tests when in a unit with a low average high school fitness score.

Cadets presumably emulated one another's workout programs. As a group, they either drilled hard or drilled relatively little. It is likely that external forces were at play and not competition, leadership, or other group dynamics, for the influence of others was mostly in one direction: having less fit companions pulled down test scores, whereas having more fit peers did not pull up scores to the same extent. Social forces arose in the opportunities to join less fit companions in sedentary activities, perhaps watching a movie or playing video games. Cadets were simply less able to join highly fit friends when they, for example, took a ten-mile run unless they were equally fit themselves.

*

Lewin understood the importance of both the person and their life context. We are not always so insightful when it comes to understanding our own behavior. We tend to underestimate how much our actions are affected by the contexts around us. Instead, we focus on our own internal decision-making. As we saw in chapter 2, our overriding belief that we are in control is called the introspection illusion.

What did you do last time you tried to change your behavior? Probably you thought about what you were doing wrong and why you wanted to change it. You focused on your desire to be successful at work, happy in marriage, or financially stable. You acted as if your desires were in charge.

A belief in free will has many advantages. It gives us confidence that we can meet life's challenges. But it also leads us to overlook the powerful influence of the physical and social worlds we inhabit. Our strong intentions blind us to the friction in our everyday surroundings—how they make some actions easy and others more difficult. The belief that our intentional self is in charge can lead us into self-delusion: we almost forget that we have bodies, and that our bodies exist in space immersed in and influenced by everyday contexts. You can forget that your self is a whole lot larger than just the smart stuff.

To see the introspection illusion in action, consider an experiment at a Canadian university with 289 students in a work-study program.[35] All reported strong intentions to save money during the term. The average goal was more than $5,000, about a third of their total pay.

Just before the study began, the students were offered a program to make it easier to save by tracking their budget. After hearing how it worked, participants judged whether it would help them reach their goals. Unanimously, they said no, it would not. This was not skepticism about the usefulness of the program. Students said that it would help others, just not themselves.

Despite their doubts, some students were enrolled by the researchers in the budget program. By the end of the term, 68 percent in the program met their goals. Only 57 percent of those in a non-budget group, who went it alone, met their savings goals. Although this difference might not seem large, it could signal disaster for students trying to put themselves through school. By overweighting the power of their initial plans, these students opted out of a beneficial program.

Underestimating the impact of our environments extends beyond novel savings programs. We fall prey to this bias even when forces are strong and clear, as shown by Stanford students randomly paired up in a study to play a knowledge game.[36] One student, the interviewer, was randomly chosen to be the "questioner," who asked difficult questions that he or she knew how to answer but others would not, like "What do the initials W. H. in [the poet] W. H. Auden's name stand for?" and "What is the longest glacier in the world?" The other student, randomly assigned to be the contestant, tried to answer. On average, the contestants managed correct responses to only 4 of 10 of these difficult questions.

The situation clearly favored the interviewers and made them look smart. The contestant was at a serious disadvantage, trying to answer questions that drew on the interviewer's idiosyncratic knowledge. Nonetheless, the unequal roles influenced self-assessments.

At the end of the study, when participants rated their own and their partner's general knowledge, interviewers believed they were better informed than contestants. Surprisingly, the contestants especially suffered from this bias. Contestants had tried to answer the questions and failed. Their conscious experience was, well, feeling dumb. Their explanations drew on these internal feelings, and they did not give enough weight to how much the rules of the game favored their partner and disadvantaged them. Contestants could easily have spared themselves, given that the questions were based on the interviewer's personal knowledge of the world. But they didn't. They did not credit the obvious external forces in this highly unequal situation and instead felt unintelligent.

We tend to overlook the influence of our surroundings, even as we are responding to them in our behavior and self-assessments. No surprise, then, that when we try to change, our go-to approach is willpower and motivation. We don't realize how much our actions are driven by our surroundings and the pressures on us. But our habits do.

Instead of beating yourself up when you fail to will yourself healthier, or wealthier, or wiser—rearrange your kitchen. Get a fruit bowl. Make it prominent. Walk to the office the slightly longer way that avoids the coffee shop with the 20-ounce Frappuccinos. Avoid your coworker who brings brownies. Forgive yourself, first, and then start making your life easier by addressing the contexts in which you live. There's no handicapping in habit formation; challenge is not the point. No pride from forming habits in the teeth of resistance. Remove the friction, set the right driving forces, and let the good habits roll into your life.

7

Repetition

A baseball swing is a very finely tuned instrument. It is repetition, and more repetition, then a little more after that.
—Reggie Jackson

You've arranged your context. You've recognized the restraining forces, the driving forces, and the pitfalls of your introspection illusion. You've turned your life into a clearinghouse for good . . . so when does the magic happen? When do the gains start? When does that second self step in and take over?

There are a couple more critical ingredients. To understand these, let's go back to some of the life challenges that might have led you to read this book.

*

Maybe it's time to get on a budget. Financial solvency seems otherwise out of reach. Yesterday, you got a follow-up notice from a credit card company. You were sure that you paid the minimum owed, but no, you didn't. You notice that you are going backward in paying off your balance. It's getting bigger, not smaller. And then there's that medical bill from breaking your wrist last year that you've almost—but not quite—paid off. The hospital has threatened a couple of times to refer your case to a collection agency.

And what happened to your New Year's resolution to join your company's retirement-savings plan? You haven't done it yet. Joining means you'll have an additional deduction from your paycheck, and you'll have to live on even less than you do now. Money just seems to trickle away: a $6 coffee here, a $15 lunch there. One night out with some of your friends can cost you as much as $100.

So it's time to take charge of your finances and figure out how to put aside money for emergencies, like a medical bill or new tires. You want to pay off those credit card balances and start saving for retirement.

At first, it feels exciting. You're proud of your new sense of responsibility. You bring your own mug to work so that you can drink the company's coffee. Six dollars saved each morning!

You bring a sack lunch and eat in the company lunchroom. But you soon decide that peanut butter sandwiches are depressing. And you miss joining the office crowd eating lunch out.

You stop by the grocery store on the way home and try to plan out your meals. Ham and Swiss cheese is an improvement, but you forget to buy mustard. So the next day it's bad coffee and a dry sandwich.

When the weekend comes, you find a free outdoor movie. But no one wants to go with you, because they've seen it already. Are you going to have to find a whole new set of friends to hang out with, given your new budget? You feel like a social outcast.

When does habit take over so this becomes less painful? When does saving money stop being a grind of austerity and self-denial and become automatic?

When does the magic happen?

*

Or maybe your challenge is to do something about family dinnertime. As your kids are getting older, you hear less about their lives. You're going to set up a regular schedule when you can all eat together. Check the calendars. Find a few nights when everyone can be there, catch up, and share their day.

You decide that, to make it possible to talk, there will be no distractions: All cell phones off. No TV in the background. No eating and running.

The first time, it's hard work. You have to corral them all to the table when dinner is ready. No one likes having their cell phones off. Your partner is only sort of on board with this new plan, and your kids stare at you resentfully.

That first dinner actually ends up making family life less pleasant. You've got a sullen kid or two and a puzzled spouse. The only real conversation is "Other kids' parents don't make them do this!" Okay, you were not expecting an immediate transition to the Waltons, but this is no fun at all.

Nevertheless, you remain committed, bolstered by research on the benefits of family meals. Kids who regularly communicate with their parents at mealtimes engage in fewer risky behaviors, do better at school, and are less likely to be overweight.[1] Of course, you don't know that you can make these things happen by instituting family meals, but the research spurs your motivation.

You persist to dinner #2. Children are still not talking, and you feel stressed. You share what you think are fascinating topics (that you carefully curated from NPR that morning). No response to your brilliant conversational skills.

Dinner #3 was hard to arrange because you had to find a night that worked with everyone's schedule. There was a mad rush to get all the food on the table at the same time. Once everyone's finally seated, you see mutinous, angry faces all around. This is getting dark.

It takes all the determination you can muster to make dinner #4 happen. The kids have found ways to communicate among themselves that exclude parents. You have to keep reminding yourself why you are doing this.

Your wonderful plan does not seem to be getting much easier. Yes, every once in a while a child joins in the conversation, but then clams up again in protest. You are getting no help from anyone.

When does the *magic* happen?

*

When do you fall asleep?

I don't mean what time you go to bed or what time you'd like to. I mean—when exactly do you slip into sleep? Did you ever try to test that when you were a kid? Did you get into bed, start to doze off, and then ask yourself "Am I asleep?" Of course you woke right up.

The truth is, it's impossible to know. First you get into bed, then you get sleepy . . . and then the sun is shining and it's time to get up.

Habit—our particular magic—works the same way. You're going to start doing those dinners or saving that money every week, and you're going to keep doing it . . . until *you* aren't doing it anymore. Your second self is, and you realize ten years have passed, and you're hearing your eldest describe to his fiancée how family dinners were always a tradition in your household. How lovely!

The magic begins silently, and you won't realize when it kicks in. You have to trust that it will happen, because it is the standard way that repeatedly rewarded actions restructure the way information is stored in our brains. Before then, it's going to be some work. Until we have laid down a habit in neural networks and memory systems, we must willfully decide to repeat a new action again and again, even when it's a struggle. At some point, it becomes second nature, and we can sit back and let autopilot drive.

But how many times do we have to repeat an action before it becomes automatic? Maybe you've heard that it takes twenty-one days of doing something to turn it into a habit. That would mean only three weeks of forcing your family to eat dinner together before they talk to one another. Only twenty-one mornings of planning your daily budget before your frugality becomes automatic.

This is a myth. The number seems to come from self-help guru Maxwell Maltz's speculation in his 1960 bestselling book, *Psycho-Cybernetics*.[2] He was guessing how long it took people to adjust to self-changes such as plastic surgery. This is a concept with much longevity but little truth.

Research provides better insight. Pippa Lally, a postdoctoral researcher in my lab, tested how many times actions have to be repeated until they *feel* automatic. She paid ninety-six students at the University of London about $40 each to participate in a three-month-long study.[3] Each student named a healthy behavior that they were not currently doing but wanted to perform regularly. Then they chose some daily event in their lives to which they could connect the new behavior. One student decided to eat a piece of fruit every day at lunch. Another chose to run for fifteen minutes right before dinner. Still another determined to drink a bottle of water with lunch.

At the end of each day, participants logged onto the study website and reported whether or not they had performed as planned. Students also indicated how automatic the behavior felt to them—to what extent did they do it "automatically," "without thinking," and "by starting before I realize."

At the beginning of the study, students gave very low ratings, about 3 points on the automaticity scale (which ranged from 0 to 42 points). They were learning a new behavior, and it did not feel automatic. As you'd expect, the more the actions were repeated, the more they felt automated. Automaticity increased the most during the first few weeks of repetition. The third time participants performed an action, they might have gone up a full point on the scale; the fortieth time they repeated it, maybe only half a point. When the action is hardest to do, right at the beginning, your habit memory is learning the most.

As a side note, it is ironic that many of the students could not stick with the program long enough to provide information on habit formation. This shows how tough it is to repeat even a simple new behavior every day. Fourteen of the 96 dropped out altogether. The remaining 82 logged in only about half the days, on average. A new action is difficult to sustain when the *only* driving forces are internal motivators of (a) wanting to do it, (b) knowing it's good for you, and (c) wanting to get the study payment. The students did not arrange external forces to drive them forward and maintain the action, forces promoting—for example—an evening jog (walk the dog, pick up mail) or eating more fruit (go to a lunch counter serving fruit as a standard side dish).

Most reassuring for those of us trying to form a new habit is that participants could miss a day or two without derailing what they had started. The day they began again, automaticity was almost as high as when they lapsed. Occasional gaps did not erase the emerging habit.

This is a crucial point. You can miss a day or two and you will not be set back to zero. An omission is not a license to cheat or keep failing. Your habit-in-formation is not so fragile that it requires perfection. It requires persistence, repetition, and those savvy context-manipulation tricks from the previous chapter. If you miss a day or fall off the wagon, don't despair. Instead, use it as an opportunity to make your context tighter, stronger, and clearer. Your habit is still forming.

In the habit-formation study, different behaviors required differing amounts of repetition to become automatic. For eating something healthy, participants had to repeat the action for about sixty-five days before they did it mostly without thinking. Having a healthy drink took slightly less repetition, some fifty-nine days. Exercise, however, required more like ninety-one days of repeated action to become largely habitual.

It's sort of obvious that some actions take longer to automate than others. If you were learning to play the piano, you'd expect a Chopin concerto to require more practice than "Twinkle, Twinkle, Little Star." We learn simple behaviors faster than more complex ones. Actions with multiple components, like getting to the gym and working out, might be particularly difficult habits to pick up.

Ratings of how automatic an action *feels* provide only one answer to our question. On average, it took participants sixty-six days of repeating a simple health behavior until they experienced it as automatic. Adopt a new behavior, do it repeatedly for two months and a week, and you've significantly increased that automated feeling.

There are other ways to test how long it takes to form a habit. Instead of asking how an action feels, we can ask what cognitive processes drive action: When does decision-making drop out so that we are not acting intentionally? One answer comes from a study of 2,228 Canadian blood donors.[4] Blood donation in Quebec is highly structured. Donors get a call from Héma-Québec when there's a local blood drive, and they just show up at the usual donation spot. Quebec thus establishes *driving forces* to perpetuate donations by scheduling and encouraging people's participation.

The study participants were selected from those who gave blood during the week of April 21–26, 2003. So all had donated at least once, with an average of fourteen times in the past, ranging from zero to ninety-seven times. Participants reported their intentions to donate during the next six months. The researchers then tracked what they actually did the following year.

As expected, beginning donors were intentional, giving blood if they had strong plans to do so, and not giving blood if they said they were less inclined to. This was true for those who had completed fewer than twenty donations in the past. More than twenty times as a donor, and their actions became less intentional, with each additional past donation incrementally reducing the decision-making involved in giving blood again. For the group who had donated forty times or more in the past, intentions had essentially no impact—they just kept giving blood regardless of what they planned to do.

Again, there was no clear line separating habits and non-habits. Instead, habits seemed to gradually develop and take over to bypass conscious decisions. The more often participants had donated in the past, the more likely they were to just show up without consulting their intentions.

For those of us wanting to form a habit, forty repetitions is a more optimistic answer than sixty-six. These different estimates come from very different settings, actions, and measures of habit formation, so there is no right number. But note that the lower estimate comes from a blood-donation setting with strong driving forces making it easier to repeat the action in a routine way. An implication is that you can lower your magic number by establishing forces that push you to repeat in the same way each time. With bigger, louder cues, your habit potentially matures faster.

Making repetition happen, however, is not easy. As Professor Chen, former head of economic research for Uber, commented, "The median driver doesn't last past ten trips. It's difficult to get drivers to stick. That's always been the primary cost [to Uber]. There's just many fewer people who are willing to drive their own vehicle than there are people to take a ride somewhere. It costs a lot of money to attract a driver. You've got to run a background check. Have a mechanic inspect their car. Do a bunch of stuff. So I invest $1,000 in you as a driver, and you do only eight trips. I've just lost a tremendous amount of money."[5]

Many Uber drivers apparently don't understand the *restraining forces* when they take this job. "The question is," Chen said, "what's the barrier? Early on, it's a difficult task. It's socially awkward. A stranger is suddenly sitting in the back of your car and you've got to figure out how to navigate that whole pick-up, drop-off relationship."

So Uber changed up the environment by adding external forces to keep drivers working. "One approach is engineering continuous pickups," Chen said. "You probably noticed this, but before your Uber driver drops you off, they often have the next trip lined up. It's just like Netflix. Automatic. Uber does a tremendous amount of analytics to chain a bunch of trips together. That's good for a number of reasons. It completely eliminates downtime, so drivers make more money. It also is automatic: 'Of course I'm about to drop you off, yeah, fantastic, now pick up the next guy.' Suddenly it's two hours later, and the driver has to actively say, 'Stop sending me trips, I need to take a bathroom break.'" In addition, Chen noted, "we spent a lot of time engineering the driver portal to avoid downtime. We only make money when the driver makes money. You don't want to give them any time to switch to Lyft or to stop driving for the day."

In Uber's case, the external forces seemed to get their magic number down to ten. That's a lot smaller than sixty-six. It just goes to show what's possible when you have the brightest minds engineering habit formation. But really, aren't you the world's leading expert on your life? Surely you know how best to cue up and trigger those family dinners and frugal spending. Your magic number is likely to fall with each piece of context that you engineer for yourself.

One month of decreasingly awkward dinners is worth an enduring, enriching family tradition. Or it could be a month of feeling slightly less deprived each time you decide not to waste your money on something you don't really need.

*

A wholesome new habit has to reckon with something else, though. Very little of our lives is a blank space, uncluttered by minor or incipient habits already. By the time we're adults, almost all of our day—and how we fill it—is the result of a squabble of contradictory habits, happening just under the surface of our consciousness.

Initially, you dreamed about the great discussions you would have with your family over dinner, and the lasting closeness you would create among the people you love most. Or you felt pride imagining your growing net worth and the satisfaction of being able to pay off your credit cards in full every month. But then reality intrudes, and your feelings begin to change. Those sullen looks night after night from your kids or the constant pangs of retail denial at your favorite stores weaken your resolve. You are no longer so enthusiastic about the benefits of your bold decisions. Your first thought is no longer "I need to make this change." It's now "This can't be worth it."

Along with the challenges of adopting new behaviors, you still have to fight off the old ones. The previous habit—the one you wanted to change—didn't immediately vanish with your decision to do something better for your family or your finances. It surfaced again as soon as your willpower started to flag. You began sliding back to where you started.

This is where repetition of the new action becomes especially useful as a tool (and not just an inert description of what habits look like). After a while, conflict resolution starts to favor the new behavior. As chapter 3 explained, processing speed is the reason.

Habits come to mind quickly. You just have to perceive the context, and the response is automatically triggered. Making a decision, by contrast, takes a bit longer and requires more cognitive control and effort. And decision-making is especially difficult when you are ambivalent about strong-arming your family to the table or making yet another frugal meal at home.

The speed with which habits come to mind gives them an advantage. When people act on habit, they experience less conflict with their desires to do something different.[6] We find ourselves acting before we have a chance to consider whether this is what we truly want to do. The speed of habits is a boon when they are what we want to do, but a bane when they are unwanted and we are trying to control them.

Repetition, then, should be thought of not as some kind of magical primer for habits, but rather as a way to induce speedy mental action. The second time you do something takes less time and mental effort than the first. The third takes less than the second. And so on. This creates a favorable mental condition for a habit to come in and take over. By the tenth time (or the sixty-sixth), you're barely thinking about it at all, and presto: a habit has been created.

Showing the dynamics of speed, a study first asked Dutch students whether biking was a realistic means of getting to six locations around town.[7] All said yes—the Dutch are big cyclists—but some responded more speedily than others. Four weeks later, the students were asked how many times they had actually biked to those six places. The students who responded fastest to the questions at the beginning of the study were the ones who biked most often. Even more telling, when biking quickly came to mind, students biked regardless of whether they had said at the beginning of the study that they intended to bike, take transit, or drive a car. Dutch students with speedy thoughts of biking just hopped on their bikes without consulting their intentions. Any of them could of course decide not to bike that day and instead take a tram. But life is complicated enough, and usually it's easier to simply act on the first thought that comes to mind.

Not that the fast-accessed behaviors are always desirable. Sometimes, we need to slow down that context-response machine to suppress unwanted habits. One night, for example, a neighbor planned to join me at our local elementary school's parent-teacher conference. She lived next door to the school, and I remember watching with some amusement as she left her house, got into her car, and drove to the parking lot—despite the fact that her front door was closer to the school than where she had to park. Her habit of walking outside and getting into her car was so ingrained that she never considered other options, no matter where she was going.

Speed isn't the only factor at work. It goes hand in hand with another consequence of repetition: streamlined decision-making. We stop considering alternative actions. Most of the time, this is efficient and functional. But sometimes we streamline decisions even when we'd benefit from considering more options.

In another study involving the Dutch and their bikes, students reported how often they used their bicycles to get to class and around town.[8] Some students rode frequently, whereas others rode only occasionally. Students then indicated how they would travel from their homes to an imaginary shop in town, with the possibilities being walking, bus, bicycle, tram, and train. Before deciding, participants could click on thirty different bits of information to learn about their destination. They had never been to this place before, so it made sense that they would gather as much data as possible, including travel time, strenuousness of the ride, weather, and probability of delay.

But students who frequently biked didn't need much data to decide. They focused mostly on information about biking. They didn't seriously consider the other options before making a decision. They looked at about fourteen pieces of information total. In the end, 82 percent of them chose to take their bikes—as they typically did in real life. Students who biked less often deliberated more. They clicked on nineteen bits of information before deciding. They focused equally on all the options, exploring the advantages and disadvantages of each mode of travel. Only 50 percent of those students chose to bike.

Repetition led to a sort of tunnel vision defined by what students had done in the past. The serious bikers didn't pause over any other options. Once their first choice came to mind, they stopped thinking that much. This difference is remarkable because it showed up even when the researchers instructed participants to consider all their options and check out all the alternatives. Under these conditions, habits still streamlined decision-making by about five pieces of information. That's a savings of nearly a quarter.

This same sort of tunnel vision benefits managers in organizations of all kinds. In one study, MBAs with about six years of managerial experience were asked to imagine that they worked at a computer company ready to launch a new laptop.[9] They were given an initial prototype to judge and then asked to compare it to three others (actually equivalent in quality). Fifty percent of the participants selected their initial laptop as the best, and they judged that the executive board at the company would vote the same. Because all of the laptops were actually equivalent, the first option should have been favored only 25 percent of the time—that would be an accurate assessment. To understand why these budding managers were making such biased decisions, researchers coded their thoughts. It turned out that the more the managers-to-be thought about the initial laptop and neglected others, the more biased their judgments became in favor of it. And they weren't just going with what seemed easy. They thought that the board would agree with their judgments—a sure sign of a successfully tunnel-visioned person.

As you'll remember from the experiment described in chapter 2, most of us will tend to go to the last (identical) item we examine in a store and declare it to be the highest quality. So why do managers go with their *initial* options? Because the cognitive control needed for decision-making requires time and energy, and managers have to make many decisions. In the real world, they don't have time to consider alternatives. They have a whole lot of options to choose from in a whole lot of different contexts. It is no surprise that quick, decisive choices are part of their style of leadership. Just as when we have habits in our minds, that first option reduces the drain of evaluating alternatives.

At home, a properly cued-up context will make your initial option the very best one. It can be as simple as hiding the remote control and putting that novel you're trying to finish in a prominent place. Even my highly motivated bicycle-racing son finds it helpful to cue his evening workouts. He sets up his bike trainer in the living room in the morning so that it's the first thing he encounters when coming home from work in the evening. Thus the initial option he considers is the one that meets his goals. With a consciously arranged context, you can turn the tendency to streamline thought to your advantage. You can make your first choice the best one, always.

*

Repetition has another important effect for our purposes: it actually changes our experience of an activity, so that it seems easier. A classic study from 2005 followed ninety-four members of a newly opened gym in the U.K. for three months to determine how people stick to working out.[10] The members had paid good money to join the gym, and initially, at least, all were committed to using it.

You know what happened next: follow-through was abysmal. (The gym business depends on it!) In the study, however, 29 percent bucked this trend. These new members consistently used the gym each week for the full three months.

Who were the ones who managed to persist? *Not* the strongest willed (as measured by their initial expression of commitment)—initially, the other 71 percent were just as motivated. And *not* the ones with the most favorable attitudes toward exercise—the 71 percent liked it just as much to start with. The 29 percent stood out in another important way.

The third of people who persisted reported initially that they were in control of their exercise routine and highly capable of regular workouts. Why did they rate themselves high in *perceived behavioral control* (as psychologists call such judgments)? The study didn't tell us. But we know that internal forces weren't responsible—liking exercise and intention to go to the gym did not spur persistence. Instead, I suspect that it had to do with situational control; the kind that comes from clearing one's schedule to make it easier to get to the gym. Maybe the persistent few had reserved Monday and Wednesday lunch hours for exercise or made sure they passed by the gym after work. Actions seem easier when we set external forces to drive us like this.

As a result, the 29 percent all did one critical thing: they went regularly for at least five weeks. They were apparently forming a gym habit. They just kept going after this point, regardless of the strength of their initial intentions. This is the familiar pattern we've come to understand: with regular repetition, we stop consulting our intentions and just keep acting (although the five-week estimate for habit formation is highly optimistic!).

What's really interesting is the "downstream" effects of going to the gym regularly. By the end of the study, when the participants completed the exit questionnaire, the 29 percent reported that they felt even more capable and in control of their exercise than they did at the beginning. They had grown more confident that they could do it. Their actions felt more fluent.

For the gym members who dropped out, however, the experience of friction actually increased. It seemed to get worse and worse. In the exit interview, the unsuccessful 71 percent started to see exercise as even more challenging and difficult than they had at the beginning of the study. They reported even greater difficulty than when they started.

Even more surprising, participants who successfully formed exercise habits across the twelve weeks reported wanting to go to the gym *more*. They were able to make the gym their first choice for a few weeks—and the simple regularity of attendance increased their desire to keep going.

Perhaps this gym study sounds obvious or circular. People who go to the gym . . . end up going to the gym. But if you read this in light of what we've discussed so far, you'll see the point: Habits come from repetition. Behavior begets behavior. There isn't a further, more complicated, rare, or special ingredient. That should be wonderfully liberating. That should make you optimistic. If you just *keep* doing it, it'll start happening with more and more ease. Make it easy for yourself. No style points.

*

So as not to oversell the power of repetition, I want to add a final caveat. Most of us repeat actions into habits in order to become better people—better partners, more effective parents, healthier, more productive, and more financially solvent. Repetition can make these things happen more automatically and help us enjoy them more.

But there are some glory seekers among us. Some of us are interested in repetition as the fast track to greatness and top performance. This idea has a long history. Aristotle is reputed to have said, "We are what we repeatedly do. Excellence, then, is not an act, but a habit" (in the historian Will Durant's paraphrase from the ancient Greek).[11] Certainly it's correct that deliberate practice, or repeated activities designed to improve performance, can make us better at tasks as diverse as music, writing, and athletics. We repeat into habit as much of the skill as possible so that we leave our conscious minds free to interpret nuances in a music score, invent creative narratives, and move with athletic grace. But this paraphrase of Aristotle is not quite accurate (and perhaps not what he meant).

Excellence and repetition are not the same. We know this from experience. We all have seen plenty of people committed to something that they just do not do particularly well. Perhaps they are in it for the mindfulness. Or perhaps they have deluded themselves effectively, and perhaps we politely assist their delusion; but excellence? No.

We all know that repetition is *necessary* to excel, but it's less clear that it's *sufficient*. Popular science has taken a strong position: Malcolm Gladwell's *10,000-hour rule* even gives a precise number. With that much practice, he argued, most of us can excel.[12] Stephen Curry, one of the highest-profile NBA players and one of the greatest shooters, seems to epitomize this rule.[13] He did not have natural physical advantages, being small and scrawny in high school, and he lacked the upper body strength to shoot the ball properly. He admitted in a news interview, "No college coaches from higher Division 1 schools wanted to recruit me and offer me any scholarships."[14] But Curry persisted and became known for his exceptional practice habits as well as his skill.[15] He might be a walking advertisement for the 10,000-hour rule. Then again, he might be that one-in-a-million person born with excellence just waiting to be tapped by discipline. Your own untapped greatness notwithstanding, research favors the latter interpretation (especially given that Curry's father was an award-winning pro player).

A systematic review of eighty-eight studies examined how closely deliberate practice was tied to performance success in music, games, sports, education, and professions.[16] With more practice, people did better at games, music, and sports, but still 75 percent or more of their success or failure was due to factors such as native talent, opportunity, and having great trainers. In education and the professions, practice made even less of a difference. However, as you'd expect, deliberate practice benefited scripted, habitual activities in all fields (e.g., editorial proofing) more than it did less scripted activities (creative writing). Clearly, it's too much to hope that simple repetition will make all of us stars.

However, part of the promise of learning about habits is that it will free up parts of your life that you previously tied up in consciously willing things you didn't need to. You'll be able to consign substantial parts of your day to your habitual self.

What you choose to do with all that free time and energy . . . that's up to you. Perhaps you'll use it to watch a Steph Curry game tape and work on your shot. Perhaps you, too, are that one-in-a-million person. With a more habit-friendly life, at least you'll have more time to find out.

8

Reward

I never did a day's work in my life. It was all fun.　　—Thomas Edison

One irreconcilable difference between you and a computer is that your patience will run out somewhat sooner than that of a piece of semi-conducting silicon. Mine, too. A piece of software will never get tired of doing the same thing as many times as you prompt it. Infinity, for a computer, is limited only by its power supply. For a machine, repetition is essentially the same as doing nothing.

But not for you. You get tired of doing the same things. You are curious. You desire diversity and stimulation. You need something more to life's routine than its inevitability.

This "something more" is the last of the three features to take into account while forming a habit. Context will smooth the way, and repetition will jump-start the engine, but if you aren't getting even a minor *reward* for your initial effort along the way, you won't get that habit to start operating on its own.

Rewards are not confusing. We're familiar with the deal from day one: we do something we wouldn't otherwise spontaneously be doing on our own in order to get something in return. If that something feels good enough, then the initial effort is worth it. But, as with other parts of habit formation, what is ostensibly straightforward has real complexity behind it.

*

Rewards, to have a role in habit formation, have to be bigger and better than what you would normally experience. That's likely going to take some forethought and creativity. It might require some deliberation on your part. Even though it doesn't sound romantic, if you want to kick-start a new habit of intimacy with your partner, you have to plan a surprising, genuine show of affection that is more than the standard peck on the cheek as you return from work in the evening. The utility of this unexpected reward is exactly that it is, well, unexpected. The size of the reward implicitly communicates that your partner's expectations have been too low. It's an invitation to recalibrate the warmth and support they can expect to get from you by sharing stories of their day over dinner, laughing at your jokes, or whatever response you are trying to build into a habit in your relationship. That feeling is the best possible starting point for new habit formation.

Here's how it works: the additional affection is unexpected, so your partner's anticipation of how you typically act was in some sense in error (called *reward prediction error*). In the brain, unexpected rewards spur the release of dopamine. Dopamine is a neurotransmitter, or chemical signal, that facilitates the passage of information from one neuron to another. When dopamine is released from one neuron into a *synapse* (the gap between neurons), it's picked up by receptors in a receiving neuron. The transmission occurs along predictable channels, or pathways, in our brains. Habit formation involves several dopamine pathways, especially the *sensorimotor pathway*, in which dopamine released by midbrain neurons is picked up by receptors in the putamen, which is linked with motor and sensory areas (sensorimotor cortices, *motor pallidum*).[1] The bigger an unexpected reward, the more dopamine gets released (along with other chemicals), and the more efficient synapses in that pathway become in sending and receiving a signal.[2]

Your partner's brain registers your unexpectedly rewarding show of affection with a release of dopamine. It then sets up the neural bases for habit formation, as neurons, synapses, and pathways work together to record and respond to what just happened. The dopamine is like a teaching signal that instructs neural areas involved in action selection to favor recounting daily events or laughing at your jokes when sensory areas encounter those same context features (namely, you at the dinner table). The signal from dopamine neurons stamps the details of the rewarding experience into memory.[3] Your partner's brain is now a bit different. It is ready to receive and acknowledge and process more affection from you in the future. You could say that you've helped your partner's brain to be more hopeful, more optimistic, more prepared for love.

Your partner is learning that sharing feelings at dinner and laughing at your attempts at humor gets affection that would not otherwise have been forthcoming. Regardless of whether your partner is the self-disclosing type, or whether you are genuinely funny, this reward is likely to encourage the behavior again. Genuine affection from you will, with sufficient repetition, build associations for them between dinnertime and intimate disclosures and between your jokes and appreciative laughter. This is an important way people strengthen their relationships. Mutual habit formation develops when both parties are big pieces of life context for each other. It might sound dehumanizing to talk about it in these terms, but it doesn't have to be. Your second self is interacting with your partner's second self at all times, just as much as your intention and will are wrapped up and entangled with your partner's. You have the power to let all these parts enable and support the others.

Unexpected rewards work in all parts of our lives, even at the grocery store. Getting the loyal-shopper discount price on milk won't change your shopping habits. But a shopping trip with an additional, unexpected daily special activates dopamine and, if repeated, could start you forming a habit of purchasing that brand. Habit formation also influences dopamine release in other areas of the brain. As your shopping habit develops, other, decision-making areas can become less active, especially the prefrontal cortex (in particular, the *orbitofrontal cortex*). After repetition, you just automatically pick up the milk without consulting that day's price. You are no longer making a decision.

Dopamine also helps us learn from our mistakes. When we act in ways that don't get us the rewards we were expecting, dopamine neurons decrease activity, signaling to avoid that action in the future.[4] Our brains respond if we come home late and miss our spouse's kiss, or if the shopping discounts end and we have to pay the full price.

This is the dark side of interpersonal rewards. Withholding affection and reacting in hurtful ways to your partner is a marker of emotionally abusive relationships.[5] When partners are not genuinely affectionate or use affection strategically to manipulate, then abuse happens. Like addictions, which we will discuss in chapter 13, such abusive relationships can be unfortunate, sometimes tragic, distortions of our normal responses to affection and reward.

Dopamine is sometimes called the feel-good chemical because it is involved in our experience of rewards. But the specific information conveyed by dopamine release depends on the timing and the relevant sending neurons and receptors. Dopamine effects play out over a timescale of seconds, with early stages of processing signaling *salience*, or something we should pay attention to.[6] Novelty and physical salience activate dopamine neurons in this way, as that unexpected whiff of heavy, sweet cinnamon buns at the airport kiosk startles you to attention. With continued processing, dopamine signals the rewards that build habits and energizes and invigorates us to pursue actions that have positive consequences and meet our goals.

All of this means one big thing for our purposes: dopamine sets a timescale to habit learning. It spikes in our brains immediately with a reward, responding to the salience and value of what we just received. Although science is still discovering much about neuronal timing, dopamine seems to promote habit learning *for less than a minute*.[7] Unanticipated rewards in the future, such as a paycheck bonus in two weeks or an athletic trophy you get at the end of a season, will not change neural connections in the same way. Rewards have to be experienced right after we do something in order to build habit associations (context-response) in memory.

Given this timing, the most effective habit-building rewards are often *intrinsic* to a behavior, or a part of the action itself. This could be the feeling of pleasure you get when you read an engaging story to your kids and see their enjoyment; or maybe the warm glow of generosity you experience when doing a good deed, like volunteering at the soup kitchen. You aren't a rat. If you volunteer, don't then go and buy yourself a big chocolate bar and expect the habit to start forming. Let the warmth intrinsic to the activity be the reward. Take advantage of your built-in humanity.

The "Fun Theory" (Volkswagen's public service campaign) illustrates intrinsic rewards in action. One project replaced the regular stairs in an Oslo subway with stairs that played like a piano when used.[8] No surprise, commuters swarmed up them. Another project programmed trash cans in a public park to make an echoing sound like something falling down a well, which actually got passersby to collect extra trash to throw away so that they could hear it again.[9]

To assess the advantages of intrinsic rewards, a study examined exercise habits among college students.[10] As you'd expect, those who liked to exercise—who rated it a fun activity that made them feel good—exercised more often and reported that it was more habitual and automatic. They didn't have to think much before heading out to the track or gym. Most interesting is that students who exercised just as often, but who indicated that they went mostly out of guilt or to please others, failed to form a robust habit. As we saw in the previous chapter, repetition is necessary for habits to form. However, repetition alone is not sufficient. Students who did *not* experience the rewards that create automaticity from repetition had to keep consciously making themselves get to the track or gym, without a helpful habit taking over. Just a slight change would help them get more out of their gym experience. They should keep doing what they're doing—but without guilt or obligation to others. By focusing on what they wanted, they would give the intrinsic reward space to come out and be felt.

Lab studies show under controlled conditions that rewards have this power. As mentioned in chapter 5, college students in one study played a computerized game of repeatedly choosing, and getting to eat, baby carrots.[11] All students indicated that they liked carrots. Some students also indicated a strong desire to be healthy and thin, which were additional rewards. These students formed especially strong carrot-choice habits—ones that persisted even when, at the end of the study, they were given the option of choosing M&M's. The more and stronger the rewards that students reported for choosing carrots, the more the repeated choice turned into strong habits that maintained despite chocolate temptations.

Rewards also can be *extrinsic*, meaning that they are not a built-in part of a behavior. Some extrinsic rewards are pretty immediate. If you are organizing family dinners to please your spouse, then his or her comment of appreciation on sitting down at the table is an instant extrinsic reward. Another extrinsic reward is surroundings that are appealing. Some gyms have upscale lobbies meant to make us feel like we are in an exclusive club when we work out. Others sell fancy exercise apparel for you to wear. These are immediate extrinsic rewards for getting a workout. They play with your feeling of class and superiority. Who doesn't like to feel special?

Of course, payment for an activity is the classic extrinsic reward. This is the kind of reward that organizes whole careers, whole lives, and whole societies. It's crude but effective. It can be given immediately, as you do something, or you might think of it during performance, but more often it becomes salient only after a delay, such as your paycheck at the end of two weeks or a month. The time lag between action and reward, along with the regular amount you receive, means that dopamine cannot do its work.

There's another reason to question the usefulness of extrinsic rewards. They *crowd out*, or undermine, our sense that we are acting for any other reason. When we get paid to do a task, we can feel like it's not something we would otherwise do. If the payment ends, we might just quit, too.

In practice, most rewards blend the intrinsic and the extrinsic. You might stay late one night at the office because you want to do your best work on a project (intrinsic), but also because you keep thinking about the recognition you'll get the next day from your boss (extrinsic).

The delay between action and reward could explain the limited success of interventions that pay people to be healthier. Health care programs sometimes offer people money to quit smoking, lose weight, exercise, or meditate. Consistent with the laws of economics, we'll do most of these when paid enough, at least initially.[12]

Consider a six-month weight-loss program with twenty-seven women and four men.[13] Their average initial weight was 209 pounds. Weigh-ins were once a month. If they showed up four pounds lighter than at the previous month's weigh-in, then they got $100. The money transferred automatically to their bank accounts. This large incentive did not yield large results. At the end of six months, participants had lost a total of about five pounds on average.

The payment did have some effect. The paid group did better than the thirty-two participants in a control group who got no money for losing weight. Control participants went through the same monthly weighing procedure and learned whether they had met their own, personal goal for weight loss that month. Across the six months, they lost only about a pound.

Three months after the end of the study, everyone was weighed again. Now the paid group had regained some of the weight they lost. They weighed only two pounds less than when they started, nearly the same as the one-pound loss by the unpaid participants.

What happened? This was a highly ambitious study, exemplary in many ways. Tracking people for nine months of weigh-ins was a major undertaking. But the program failed to form healthy eating habits. If you think about what we've already learned about habit formation, it's easy to locate the problems: repetition and reward (perhaps context, too, but that's less clear).

There probably wasn't much repetition at all in this program. My guess is that participants started off each month not thinking much about weight loss. As the weigh-in day neared, they started dieting. They might even have fasted the day before weigh-in. After all, $100 is a lot of money. By escalating a diet in this way, participants weren't repeating new eating habits. To our conscious minds, such repetition seems superfluous. It shouldn't matter if we are sporadically starving ourselves and sporadically ignoring our diet. We assume that a calorie deficit is all we need. But if we want to form habits, we need to repeat actions enough so that they become automated.

The reward also wasn't optimal. It was given at the end of the month, and it wasn't closely tied to performing any specific behavior. Maybe participants thought about it sometimes when they were trying to diet. But the rest of the time, the reward could not cement in mental connections between contexts and responses. As a result, new habits didn't form and new behaviors didn't stick.[14] To our conscious minds (and to plenty of economists), big rewards should work. It seems highly motivating to win $100 for monthly weight loss or to treat yourself to concert tickets for meeting a work deadline this week. But it's just not habit-forming. The reward is not tied closely enough to your behavior. Single large rewards are not designed to build habits.

Employee wellness programs offered by many businesses in the United States fall short in many of these ways in their attempts to form new habits such as losing weight or not smoking.[15] Rewards include lowered insurance premiums and sometimes cash payments over time. Few such programs teach people to repeat specific actions. So there is little habit formation here.

You might wonder about negative rewards, or so-called contingency contracts. You agree to some aversive event (paying money) that you can avoid by doing something (losing weight). A variant is the "swear jar" in many homes. If someone swears, then they are punished by paying, say, a dollar to the jar. This, along with the inevitable family ridicule, is presumably enough to reduce the offender's dopamine response and decrease their swearing. In this example, the behavior is tied to an immediate consequence (at least when others are in hearing range).

But more often, contingency contracts are set up in ways that are not optimal to maintain change over time. Maybe you bet your brother $100 that you'll pass your state's bar exam the first time. If you flunk, it'll cost you. So you hope that this threat will create a new study habit. Or maybe you decide that you need to get to the gym. If you don't go three times per week this month, you can't buy that jacket you covet.

These might be effective motivations in the short term. But they're not the kind of rewards that form new habits. They are too far removed from the behavior you are trying to change, and they are not necessarily tied to any specific repetition.

Given the way that dopamine works to create habit associations in memory, immediate rewards for lots of repetitions are key.

*

There's more to the dopamine story than immediacy. As we discussed, dopamine responds to uncertainty in the form of reward prediction errors, which enable us to learn from the experience. This means that we learn from unusual or unexpected rewards—bigger than or different from what we're used to. This might be the most surprising idea so far.

Have you ever managed anyone? If so, have you ever heard the advice that it's critical that you carefully explain your expectations for their work, and what the rewards will be? The workplace wisdom is clear: Rewards (or remuneration) should be transparent, reliable, and firm. Surprises are out. Predictability is in. That's how to get the most out of your employees—and yourself. Probably you know exactly how much is in your own paycheck each period.

The workplace wisdom builds trust and reduces confusion and stress. But it's not the way to efficiently build new habits. Habits depend on surprise. Yes—the most boring and repetitive of our behaviors actually depend on our being disrupted and a bit off-balance. And it all has to do with this third and final feature: *uncertain rewards* matter most.

Uncertainty of rewards lures us to casinos. Nearly 70 percent of gaming profits now come from electronic slots and video poker.[16] Machines are programmed to display near-misses more often than chance, heightening players' feelings of "I almost won!" Getting so close to winning feels like an accomplishment, which can activate dopamine reward pathways and strengthen habits that keep us in the game (see the discussion of addiction in chapter 13).

Why would this be so? An evolutionary explanation is that all animals are sensitive to uncertain rewards because, in the wild, repeated foraging in the face of scarcity was required for our survival: if we were going to find food, water, and mating opportunities, we had to persist despite repeated failures.[17] Dopamine, then, may motivate us to keep trying despite infrequent success.

We are all pawns of uncertain reinforcement. This gets clearer when we think outside the context of the workplace. When was the last time you checked your phone? Americans check them 8 billion times a day, which means an average of 46 times per person.[18]

Smartphone use is highly habitual. One trigger is time of day. For many people it's the first thing they do in the morning, before they even get out of bed: Wake up—check phone. It's the last thing many people do at night: Get into bed—check phone. During the day, many people check their phones when they succumb to ennui: Feel bored—check phone. The reward for all of this phone activity? Every so often, an email, text, post, or tweet is interesting. Most of the information is an irrelevant time-waster. That one nugget of useful information or juicy communication, that single occasional reward, keeps us checking regularly.

Animal research has clearly demonstrated the power of uncertain rewards. In one study, mice pressed a lever for a food pellet. This reward was given at random time intervals. Sometimes pressing after nine seconds won a pellet, and sometimes waiting thirty seconds was required.[19] This intermittency is similar to that of some natural rewards. A bee collecting pollen from a flower has to wait before going back to that particular blossom, so that more pollen can be produced. Sometimes the wait is long, and sometimes it's short.

When rewards were given at random intervals like this, mice ended up pressing the lever a number of times without getting any food. They couldn't know which press would actually deliver, so they kept at it. They formed a strong habit for lever pressing that persisted even when the rewards completely stopped. In a workplace or at the gym, that's called productivity.

To our conscious minds, larger rewards and more certain rewards—ones that we know are coming—are motivating. But habits thrive on uncertainty. Imagine that you are participating in an auction that involves chocolate coins as a reward. You can bid on a lot containing five coins or on a mystery lot that contains either three or five coins—you won't know which until after your bid is accepted. Logically, the lot with five coins is worth more.

But it wasn't. Researchers at the University of Chicago staged just this auction and found that the average bid for the guaranteed five-coin lots was $1.25. The average bid for the mystery lot was $1.89.[20] When asked, participants said the uncertain auction was more exciting. It didn't increase the actual value of the reward. It just made the game more fun. Participants paid more to play and said they wanted to participate in the auction again. (The secret, though, was getting caught up in the process. When participants planned their bid in advance, they preferred the certain reward.)

"Gamification" builds on these insights about reward. Many video games, structured with uncertain rewards, establish strong habits. In 2018, the video game industry was worth more than $130 billion.[21] Educational games also benefit from uncertainty. When college students tried to learn concepts through playing a game, a correct response won either a set number of points or a number of points depending on a roll of dice.[22] When rewards were determined by the dice (and therefore uncertain), students spent more time answering questions and were more accurate. Gamification is being used in all kinds of job-training programs. To teach skills to fighter pilots, auto mechanics, and laparoscopic surgeons, games offer many different types of rewards, including badges and point systems. However, only a few teaching games involve uncertain rewards, and, perhaps as a result, games are often no more effective than standard teaching programs.[23]

In short, uncertainty pulls on reward systems in the brain in ways that may not seem rational, but nonetheless keep us doing what we're doing.

*

Rewards are also an excellent way to measure how strong a habit has become. In the previous chapter, we saw how habits can creep up on us and get established without our being fully aware of them. That doesn't mean that we can't measure their strength.

For scientists, *insensitivity* to reward is the gold standard for identifying a habit.[24] The only way to know for sure if an action is habitual is to test what happens when the reward changes. If we persist even when we don't value the reward as much or it's no longer as available, then it's a habit.

As mentioned in chapter 3, this phenomenon was first discovered in research with lab rats. In one study, for example, rats were trained to press a lever for food pellets one hundred times or five hundred times.[25] After this initial learning, the animals were fed a few pellets and then injected with a toxin that made them ill. Rats quickly developed an aversion to the pellets. What had been a reward now was seemingly poison— the same food aversion that you and I develop from food poisoning.

After this experience, the rats trained to press only one hundred times did the logical thing: They stopped pressing the lever. They avoided getting the pellet that apparently made them sick. Rats trained to press five hundred times, however, had hit the lever often enough to have formed a habit, and even after the food became associated with feeling sick, they continued to press it. If they got a pellet and started to eat, these animals were observed to spit it out in disgust. It was clearly no longer a reward.

Habits, however, didn't keep those rats lever-pressing forever. Instead, rats' habits were modified through experience. After a few minutes of pressing for no real reward, rats seemed to figure out that the lever no longer got them anything they wanted, and they quit.

These kinds of studies reveal a very basic feature of habits: the action is cued regardless of whether it is desired at that moment. It's as if the ghosts of prior rewards stick around. The practiced action (hitting the lever) popped into mind, and rats performed it without deliberation. This shows how the effect of rewards can persist and stretch into the future. Rewards are extremely efficient in this way: they continue to operate on our habits well past the last time we got them. A well-chosen reward is like a really solid, steady investment.

My colleague David Neal and I decided to test exactly this aspect of rewards in an experiment involving everyone's favorite overpriced theater snack.[26] We went to the local cinema on campus and gave viewers popcorn to eat. Stale popcorn is unpleasant, but it won't make people sick. So we created some by popping huge vats of popcorn and letting it sit in our lab for a week.

The theater let us show a few short trailers before their feature film. The study, we announced to participants, was to learn about their movie preferences. We gave each person a bag of popcorn and a bottle of water, supposedly as compensation. Half the participants got the stale bags and half got the fresh. After viewing the trailers, participants turned in their bags with any remaining popcorn, so that we could measure how much they ate. Moviegoers also rated how often they usually ate popcorn at the movies—our gauge of habit strength.

Participants who did not report a movie-popcorn habit acted rationally and ate a lot more of the fresh than the stale popcorn. They ate 70 percent of the fresh popcorn bag, on average, and about 40 percent of the stale. This was a campus, after all, and the free food might explain why even stale popcorn got eaten. In contrast, the moviegoers with a popcorn habit ate the same amount, more than 60 percent of their bag, whether the popcorn was fresh or stale.

Later, everyone told us that they hated the stale popcorn. But that didn't stop people with habits. When in the cinema, they ate popcorn as they always did. They were totally insensitive to what we might call their current pleasure. We would expect them to be actively judging what they were consuming and making a call on whether or not to keep eating. Instead, the cues were too strong: lights down, trailers flickering, popcorn bags in hand. They acted true to form.

In a second study, we made a minor adjustment that created friction on habitual eating: we put paper handles on the bags. Half of the moviegoers were told to hold the handle with their dominant hand (usually the right) and eat using the other. Try it sometime—it's sort of like starting to use chopsticks when you typically eat with a knife and fork. The rest of the participants were told the opposite: hold the bag with your nondominant hand and use the dominant one to eat. They were essentially eating the way they always did.

Those eating with their nondominant hand could no longer just eat as usual. They had to deliberately pick up the kernels and transport them carefully. With the added friction, those with strong popcorn habits ate only 30 percent of the stale bags and 40 percent of the fresh—a significant decrease from when they ate in the normal fashion. Disrupting eating habits even this tiny bit made them think about what they were doing. Suddenly, they were acting on their actual experience in the moment—their genuine dislike of stale popcorn—not on their habit of eating popcorn in the past.

Popular media love reporting this kind of research, and ours got its fifteen minutes of fame. But they misinterpreted the results. Health magazines concluded that the popcorn-bag-handle study showed the weight-control benefits of eating with the nondominant hand. In their view, this was a way to eat less. When they contacted me for interviews, I tried to point out how this would backfire: eating with our nondominant hand seems to make us pay attention to the taste of what we are eating. Participants didn't much like even the fresh popcorn in our study, and they hated the stale. So it makes sense that they ate less when paying attention to what they were doing, even when the popcorn was fresh. But if we really like the food? When attending to experience in the now, we might eat even more than we do normally. Eating with your nondominant hand is not a dieting technique. It's a way to derail a habit of automatically eating—to be more aware of the food.

The waning of rewards' effect explains why our newfound frugality persists long after our credit card debt is paid off and the pride of saving money is a distant memory. Our behavior is now on autopilot. Even the very wealthy can get stuck with frugal habits like this. Warren Buffett, chair and CEO of Berkshire Hathaway and one of the richest people in the world, lives in the same home he bought for $31,500 in 1958. Charlie Ergen, founder and chair of Dish Network, still packs a brown-bag lunch from home every day, consisting of a sandwich and Gatorade. Hilary Swank, Lady Gaga, and Kristen Bell, all high-earning celebrities, reportedly clip coupons before shopping. Yet bad habits persist, too. A habit formed from watching rewarding seasons of *Game of*

Thrones continues even when the networks fail to produce any other shows that are similarly enthralling. Our habit self doesn't notice. It keeps us watching screens in the evening instead of switching to book or music alternatives.

*

Once you understand the way rewards work to build habits, it becomes easy. Handwashing with soap is one of the cheapest, most powerful health interventions in the developing world. How do we make this sufficiently rewarding so that children repeatedly wash their hands?

Enterprising researchers delivered child-size bars of translucent soap to four-year-old children in an impoverished community in Western Cape, South Africa.[27] For some children, the soap was rewarding—brightly colored and translucent, with a toy (ball, plastic fish) clearly visible in the center. Others got the same toy, but separated from the soap. At the beginning of the study, the children rarely washed their hands before meals or after using the toilet. After getting a new bar every two weeks for two months, children using the fun soap washed their hands more often than those with the regular soap. Handwashing became immediately rewarding, getting kids closer to the toy inside.

How about handwashing rewards for adults? The Mrembo handwashing station, designed for use in rural Kenya, has a mirror on top of a small sink.[28] When placed outside of a latrine, the station rewards users with a view of their own faces while they wash their hands. And really, what could be more rewarding than your own image?

Habits are built in the moment, from our experience of pleasure. The selection rule is simple—what we find enjoyable. In short, we learn habitually when our actions repeatedly bring us more pleasure than our neural systems expect.

9

Consistency Is for Closers

Stability is not immobility.　　—Klemens von Metternich

Your habitual self has different appetites than you do. This difference is critical when we are attempting to steer our whole selves toward our preferred behaviors. Habits, as we have seen, thrive on reward uncertainty. Beyond this, habits don't crave variety. In fact, they hate it. Variety weakens habit. Variety attenuates its power to direct your behaviors. This is because variety is the enemy of stable contexts, which, as we have learned from chapter 6, are the sine qua non of habits. If you aren't arranging your life to reliably, unfailingly cue your new desired habit, then that habit will never take hold. Only by keeping your life as consistent as possible will your habit grow. Otherwise, you can expect it to develop only slowly, like a plant with far too little light.

You and I both have kitchens, and probably you, like me, make coffee there first thing in the morning. But the usual cues in your kitchen context are different from mine. If you use a drip coffeemaker, yours are filter, grounds, water, glass carafe, and drip machine. My espresso maker has different cues: a filter basket, espresso grounds, tamper, water, espresso machine, and milk steamer. Maybe you have a kitchen island where you sit and wait for coffee to brew—another cue. I have to stand to make the espresso and foam the milk. All these are recurring components of context that facilitate both of us making coffee. With enough repetition, these cues become folded into our morning habits.

Of course, this morning your kids might have left their train set on the kitchen floor, so you stumble over it trying to reach the pot. Or you forgot to pick up filters last time you were at the store. These changes alter the cues. You are forced to suddenly think about what you're doing. Should you put the trains away or just step around them? Improvise filters from paper towels? Do you even need a cup of coffee now, or can you pick one up on the way to the office?

With a change in context cues, you have to think. You can't just act out of habit. You even have to decide how much you want coffee right now. If it's too difficult, you might decide to wait until later.

But after coffee, maybe you take a short jog. If you don't have the coffee, you don't jog. And when you jog, you typically use a phone app to track the miles. The app beeps, and you know you are done. Your phone is part of the context of your run. The beep is a cue indicating when to stop. A very nearly literal one. It makes your run almost automatic.

Overnight, however, your phone updated to a new operating system. You no longer get the welcome beep. Yes, it's a small change, but such changes in cues force you to make decisions. Is it worth going online to figure out how to update the app? Maybe you just estimate the distance this morning. The absence of that regular cue would put a hitch in your morning run.

Or perhaps the context for your morning run involves a partner. You head to your usual meeting point where she joins you. She's a human cue to speed up the pace (but you don't have to tell her that). Another trigger for your run is the time of day. If you dawdle over morning coffee, you'll miss your running partner, and you won't get home in time to shower before work. Still more cues, supporting whatever other actions you're performing: finishing your coffee, seeing your kids off on the early morning bus, lacing up your shoes. You don't go for a run until all those are completed.

Locations, electronics, people, time, and other actions: all are stable cues that become tied to exercise to make up your morning habit. Change one, and it could undo your habit and make you think, at least right then. Change one permanently, and it could eliminate the habit altogether.

In this chapter, we're going to learn how important it is to keep your habit-promoting context as stable as possible. If you set up your world to be constant, recurring, and unwavering, then cues can be the jet fuel to make your new habits take off with stupendous speed. Our minds can start to develop those context-response shortcuts that automate meeting our goals.

*

For the formation of gym habits, there is great power in the context cue of time.[1] Over twelve weeks, some new gym members in a study developed patterns of exercising at regular times of day. One reported going "every morning at 7:00 a.m.," and another went "daily after supper." Others reported that they exercised less regularly, whenever they could find time. At the end of twelve weeks, those who exercised at the same times of day reported that they did so without thinking much about it or reminding themselves to go. For them, exercise had become automated. Those who worked out at inconsistent times weren't so lucky. They seemed to have to rely on that old model that we've been trying to get rid of: they exercised only when they *wanted* to or when they consciously forced themselves to go.

Timing also matters when we need to take regular medications. Keeping up with daily medications to control blood pressure or for birth control is especially challenging because there are no disease symptoms to remind you, no helpful pain cue to trigger the habit. But missing daily doses can be disaster in both cases.

Again, timing is a key to such habits. A particularly convincing study tested the advantage of time cues for taking hypertension meds. The researchers replaced medication bottle caps with special ones that recorded how often and when patients took their meds.[2] In general, compliance was high, and about 76 percent took the pills at the prescribed time. Nonetheless, patients who had earlier reported stronger habits of taking the pills at a certain time of day were especially compliant. They took the pills more often, particularly within two hours of the prescribed time. A similar study with oral contraceptives revealed less overall compliance, with about half of the participants admitting that they missed doses each month.[3] But again, time cues mattered. Among those who missed twice or more a month, only 44 percent had a regular time to take their meds, whereas 90 percent of those who never missed a dose used time cues. It didn't matter when the women took the pills—morning, afternoon, evening, or night. Just doing it at the same time was critical.

To our conscious minds, stable cues aren't a big deal. Taking pills at different times of day shouldn't matter if you're sufficiently motivated (and what could be more motivating than your heart health?). In fact, the researchers with the fancy hypertension pill bottles expected greatest compliance among patients who believed in the efficacy of the meds. These are the people who should be most motivated to take them. But patient beliefs had no impact on repeated medication compliance.[4] Instead, stable time cues were what kept patients compliant.

These studies are good illustrations that "context" definitely does not just mean "physical environment." Location is important, but your context can also consist of intangible things: the time of day, for instance, or your state of mind. One of your most important possible contexts is other people (as you are for them).

The people around you can be stable cues, especially in close relationships. To your partner, you are a stable cue that activates certain responses. In return, your partner is a cue that activates some of your responses. He or she might text a list of groceries to you, and in this way, provide a cue to you to stop by the store and pick up stuff for dinner. Or you might get gas in the car on the way home from work, enabling your partner to pick up the kids from school, which is a cue for you to make dinner. Of course, we don't experience our relationships as cues and responses. That would be awfully unromantic. When we first start a relationship, we think about the other person's feelings and expectations for us. We don't expect potential mates to text us a grocery list, and if they do, we might have to think long and hard about what that means about them and the relationship. But once we become closer, we set up a kind of *behavioral interdependence* with our partners, so that our actions are smoothly intertwined.[5] Interconnections become increasingly strong. We come to rely on them often, for important things, and in many different ways. We are each serving as a stable cue for the other's response, and the other in turn accommodates to us.

Over time, these automated interaction sequences may be rehearsed and organized to the point that they become relatively automatic and occur outside of awareness. The automated way that each partner cues specific actions for the other explains a puzzle in relationships: How can people be very close, in successful relationships, yet have little conscious experience of that intimacy and closeness? An answer is that we do not have to be consciously aware of our habitual *meshed interaction sequences* with our partners.[6] They run off automatically, with each partner habitually facilitating and augmenting the other's actions. Successful couples thus interact in relatively mindless ways, not thinking much about what they are doing or why they are doing it. We expect our partners to keep being the rewarding, wonderful people that we have come to love. As a result, our dopamine reaction stays pretty much at neutral. Remember, by the logic of *reward prediction errors*, we react to the rewards that we are not expecting, but not so much to expected rewards.

That may sound weird—the idea that successful couples exhibit mindlessness—but think for a moment about the most intensely engaged couples you can imagine. The ones who never leave each other's side, who stare longingly into each other's eyes at all times, who are surprised and delighted by the slightest action of the other. Who does that sound like? Teenagers. Romeo and Juliet. First love. Bright and hopeful (and, we hope, not doomed).

Yes, forming expectations about our rewarding, wonderful partners has an ironic implication: successful couples may not actively experience much passion for each other.[7] It's as if they each keep bringing the other the same flowers and presents, but neither is noticing anymore. In real relationships, of course, the interdependence is more likely to involve one partner paying the bills on time while the other washes the dishes. But the point is the same. Relationships can automate so that emotions and intimacy become *latent* in the sense that partners are closely tied but do not consciously experience their passion for each other.

In fact, successful couples may experience no more intimacy on a daily basis than couples in *parallel* or *empty-shell relationships*, in which partners have little real relationship or meaningful impact on each other.[8] For those who are successful, cues and responses run so smoothly that decision-making rarely brings the relationship to mind. At best, this smooth cuing and responding is the basis for security and trust in a relationship. A potential but not inevitable downside, as we discuss in chapter 11, is boredom and taking one's partner for granted. Variety may be the enemy of your habitual self, but it's still the spice of life. Remember—you can't run on habit alone. As always, we have to remember that our habitual lives are best thought of as supporting us by liberating more of our mindfulness and attentional resources for other things.

*

Our minds are designed to miss the forest for the trees. We are primed by cues and end up not seeing the bigger picture, the world at large. Much of our life is conducted in a kind of surreal landscape of giant cues that blot out the proportions of the reality underneath.

René Magritte's fanciful painting on page 136 illustrates this feature of habit cues (*Les valuers personnelles/Personal Values*, 1952). The cues that activate our habits have outsize influence. Morning in your bedroom? The shaving brush, soap, glass, and comb loom

large. The bed dwindles in comparison. Today, this would perhaps depict your phone on your bedside table, blaring out the alarm. Time to wake up. Your mind registers nothing else (at least until coffee).

We are well aware of some things that catch our attention. When we're hungry, we find ourselves looking longingly at the hot dog stand outside the hardware store. When we are thirsty, it's hard to ignore others enjoying a cold drink. Habit cues, because they are built on our history of rewards, similarly attract notice. As discussed in chapter 8, when we get rewarded, especially when the reward is unexpected, our neural systems respond with dopamine signaling. This neurochemical helps to establish mental connections between contexts and responses, forming habits in memory. But it does more than this. Dopamine also directs our attention. It makes sure we respond to the cues that got us rewards in the past. The neural systems activated by such cues quickly send signals to influence our reactions. This is why we notice habit cues even before we make decisions about what to focus on.[9] Habit cues grab our attention faster than many other aspects of our everyday contexts.

The attention-grabbing effects of cues that got us rewards in the past were shown in a clever lab test.[10] The cues in this case were circles on a computer screen. The task was easy—find a red or green circle, among many other colored circles shown, and then hit a key to indicate whether the line inside the circle was horizontal or vertical. For some participants, the green circle gave a large reward (ten cents), while the red one got a small reward (two cents). For others, the payoffs were reversed.

College students played this game over and over, 240 times—often enough to form a habit of clicking a computer key when they saw a red or green circle. In doing so, they earned a few dollars. Eight days later, the students came back to work on a different task. This time, the color of the shapes was irrelevant. The task was to find the one shape on the screen that differed from the others, such as a triangle among the circles. This should have been simple, but it wasn't easy for everyone.

For students who got the big reward for finding green circles in the initial study, the green circles were now distracting. When one was on the screen, they could not easily complete the current task and find the odd shape. That green circle was there, grabbing their attention, slowing down their response. They seemed to see it first, before the shape they were actually seeking. The exact same thing happened when red and not green was the highly rewarded color in the first task. Now if a red circle was on the screen, students were slow to identify the odd shape. That red circle had their attention.

Logically, this shouldn't have happened. There were no rewards in the second task. The first study, with the rewards, had been eight days earlier. Cues are just that resilient.

This works outside the lab, too. Walk into your office and see a star client or prospect sitting at your desk, and your attention is immediately directed their way. You find yourself greeting them before noticing who else might be in the office. You simply do not perceive the world in an objective way. Threats loom large. So do promising leads and advantages.

There's a term in the military, especially in the air force, for computer-generated overlays on your field of vision, often projected into a transparent eye guard. They call them *heads-up displays* (HUDs). They'll show the most important stats to the pilot, for instance, without requiring a look down at the instruments. Inevitably, this technology is also starting to be implemented in cars. Many newer models now feature projections of your speed onto the windshield area itself so that you don't have to shift your field of vision by glancing down to the dashboard to check it.

Our minds do this for us, but even more invisibly. You can program the HUD in your new car. Similarly, by forming habits, you can train your mind to select for cues in the world that you have chosen, and these, too, will loom large in your view at all times.

Our minds are also sensitive to broader settings that signal what specific cues and responses get rewarded. In still another study, some students got rewarded for the green circles only when they appeared on a background of a black-and-white picture of a forest.[11] When the background was a picture of a city, the red circles got the reward. When later tested with instructions that neither the red nor green circles got a reward, the green circles were distracting only when the background was a forest. The red circles were distracting only when the background was a city. Thus the cue, red or green, captured attention only in the setting in which it had been associated with rewards in the past. In the other setting, the color hadn't been rewarded and so did not capture attention. The rigidity of habitual responding is offset, it seems, by its specificity. It adaptively orients us toward the particular cues that, in a given setting, maximize our chances of getting a reward. Thus, if Magritte's painting were depicting a shaving brush, soap, glass, and comb in our kitchens or living rooms, these items would not loom so large. They are rewarding early in the morning in the bedroom. Cues and contexts are paired in our minds in a sort of habit-inspired caricature of the real world in which we live.

Yes, the research on circles involved an abstract computerized task. Nothing at all like our everyday plans to save money by not watching the Home Shopping Network, or to work harder and more effectively at our jobs by not procrastinating and checking our Twitter feed every few hours. But this is the beauty of highly controlled lab research. We can see the effects of simple reward history, separate from other things. We learn that our attention is grabbed by even abstract, senseless contexts and cues that have been rewarded in the past. We see them faster and respond to them before we have time to think about doing something else.

A whole host of context cues in our environment has the same effect as seeing a valued client would. When we have been rewarded repeatedly for using particular objects in our environment, they automatically capture our attention. When we have a habit of saving, our attention is automatically focused on the sale racks in a clothing store and the generic labels in a supermarket. We don't pay much attention to the promotional ads that pop up when we surf the web. We are drawn to the cues that in the past generated pride and feelings of success—the cues that activated our past purchases. With this capture of attention, cues keep us repeating beneficial actions.

We are not, of course, pawns of the cues around us. If we understand the power of stable cues, then we can harness it to more easily form desired habits by controlling the contexts of our lives. Forming habits, it seems, is about establishing stable cues that support your desired actions.

*

The benefits of consistency and stability can be clearly seen in the achievements of exceptional performers. Ever wonder how musicians remember how to play long pieces of music from memory so that they can perform seamlessly in concert? It's effective memorization, of course, and years of dedicated practice. But they do more than just stare at a sheet of music when they practice. Accomplished musicians practice in a way that sets stable cues in music scores. It's similar to the way that we develop our own mental maps of the world, paying particular attention to street signs and distinctive buildings as we learn to navigate through a new city.

I talked with Dr. Tania Lisboa, professional cellist and research fellow at London's Royal College of Music, about how she learns a piece of music.[12] Dr. Lisboa explained, "Students, especially the young ones, practice [a piece] from beginning to end, beginning to end, beginning to end. Pretty automatically. When the action breaks down, students can't start up again in the middle. They can't break halfway through that sequence of actions. They have to go back to the beginning and start again." Beginners, it seems, chunk a whole piece of music together in their minds and just play the whole thing. They have no cues in the music except the start and finish. Their experience is much like you being asked for the fourth digit in your phone number. To retrieve it, you'd have to start in sequence with the first number.

Memory can fail, we humans are all too fragile, and we get distracted easily (not to mention, classical music audiences seem to cough a lot). But expert musicians don't get stuck when they hesitate or have memory lapses. They set stable cues for themselves within the score. "Experts practice a piece from beginning to end," Lisboa told me, "but also work in sections. You start and stop at parts of the music—at the beginning of a phrase going to the end of the phrase." Cues can also be expressive, at sections that are sad or happy, or perhaps changes in tempo or in bowing and fingering. "By practicing in sections, you have performance cues; points that will guide your memory of a piece. You are on automatic pilot when playing, but you have those points of reference. Those points," said Lisboa, "bring you back to the performance and to the actions you need to do to execute a piece or to project the musical idea."

Expert musicians, it seems, have learned to chunk together smaller sets of contexts and responses. Their performance doesn't suffer if other musicians make mistakes or the audience keeps coughing. Even in music, context cues are useful. They can automatically trigger playing the next phrase of music going forward.

*

There's one more technique of context consistency that is important. It flows from the idea that responses themselves can become cues . . . for additional responses. This is a bit like the chunking that musicians do, but it's practiced all around us, mostly without our realizing.

Fire-prevention associations have campaigned for years to get people to change their smoke alarm batteries when they change their clocks to daylight saving time and back again.[13] The idea is to use an existing behavior as a fire-prevention cue. We can *stack*, or *piggyback*, the behavior of replacing batteries on top of adjusting clocks. The existing behavior is a stable context—you have to do it twice a year. With practice, everything gets cued together: change clocks—replace batteries. Some fire departments give out free batteries around March and November, to encourage stacking fire-safety chores onto time-change chores.

When you repeat an activity with several components, and you do them in the same way each time, your brain connects the actions together into a unit. The whole sequence is treated as a single item in your mind.

In evidence that stacking works, consider dental floss. Many of us brush our teeth regularly but fail to floss.[14] To test whether stacking increases flossing, researchers gave fifty British participants, who flossed on average only 1.5 times per month, information encouraging them to do it more regularly.[15]

Half of the participants were told to floss *before* they brushed at night, and half *after* they brushed. Note that only half of the participants were really stacking—using an existing automated response (brushing their teeth) as a cue for a new behavior (flossing). The other half, who first flossed and then brushed, had to remember, oh, yes, first I need to floss, before I brush. No automated cue.

Each day for four weeks, participants reported by text whether they flossed the night before. At the end of the month of reminders, they all flossed about twenty-four days on average. Most interesting is what they were all doing eight months later. Those who stacked, and flossed *after* they brushed, were still doing it about eleven days a month. For them, the new behavior was maintained by the existing habit. The group originally instructed to floss *before* they brushed ended up doing it only about once a week.

As a business strategy, stacking is sometimes called *piggyback marketing*. Two different companies team up so that the carrier company's existing product becomes a cue to use a rider company's complementary one. Piggyback marketing explains how PayPal so quickly gained in popularity. It was integrated into eBay very early on. While people were making eBay purchases, they got used to seeing and using

PayPal. With enough experience, many purchasers developed a Pay-Pal habit to go with their eBay habit—and then a PayPal habit that extended way beyond eBay purchases.

This strategy helps to explain the rapid growth of many social media sites. Instagram was initially banned from Facebook, but eventually they integrated, so that Facebook became a stable cue that triggered use of Instagram features. YouTube connected to Myspace and eventually took over as the major video-posting website.

Many new businesses launch by using a kind of piggybacking strategy. Your new business venture is doing this if you are starting off by working as a freelancer for an already-established agency. The idea is to piggyback on their success in ways that allow you to automate some of the many activities that are required in a launch. You can, for example, take advantage of their marketing and access to clients while you ramp up your own skills, perfect your craft, and acquire business acumen. Then, when you are ready to evolve out of the agency-freelancer model, you can morph into your own business (while of course avoiding conflicts of interest with their clients).

Tying a new behavior to existing cues is a useful life hack for forming a new habit. The new behavior quickly becomes automated. After all, the automaticity is already in place. You just have to add a new step.

Stacking is most successful when the new behavior is compatible with an existing habit.[16] Take your meds at night? Easy to remember when you put them on your nightstand and tie taking a dose to checking your phone before you go to bed. If you head out of the office at 10:00 a.m. to get a coffee at Starbucks, make that the time you answer at least one email you've been putting off. The cues will stack, and soon enough the pain of answering a difficult email will fold into the reward of the coffee—and presto, you have a brand-new integrated habit.

Procter & Gamble engaged my lab to test how stacking works with new products. P&G provided us with spray bottles of a new fabric refresher, and college students used these for a month.[17] With one spray, they could remove smells from their clothes. But they had to remember to use it. To make it easier, some students were instructed to stack the refresher onto their existing laundry routine. For example, they might plan, "When I pick my jeans up from the floor, I will refresh them before putting them on." Or, "Instead of putting my shirt in the laundry basket to wash later, I'll refresh it, and hang it up."

At the end of every week, students reported back to us how many times they had used the new product. Students liked the new product and used it pretty often. But stacking got them to use it more—especially the students who typically thought little about their laundry and thus were the kind to forget about the refresher. With stacking, they remembered to use it thirteen times during the month. Without stacking, they used it 15 percent less often.[18]

A related strategy of building new behaviors onto existing cues involves *swapping* one behavior for another. The habit cues that automatically activated an old response can be co-opted to activate a new, similar response. Swapping explains the immediate popularity of soy milk. Without much thought, lactose-intolerant consumers started to use it as a substitute for cow milk. Tofu had a rockier start in U.S. markets. It could not easily be integrated into standard American recipes because it did not cook like animal proteins or cheese. Eventually, tofu was incorporated into ice cream and gained some popularity as a dairy substitute.

In a direct test of swapping, Jen Labrecque and I asked consumers to identify two products they had recently purchased—one they had actually used and another they hadn't.[19] The interesting question was whether each product replaced something they were already doing. For example, an e-book reader swaps easily for paper books. A Swiffer floor cleaner does away with a broom or mop. In contrast, for an aspiring exerciser, a new piece of exercise equipment did not replace anything they owned previously. As we expected, new products were more likely to be used when they completely replaced an existing product. They were seamlessly inserted into an existing habit.

Swapping is one reason that Americans' decreased consumption of sugary sodas in recent years has aligned with increased consumption of bottled water.[20] Water is sold in individual-serving bottles in convenience and grocery stores—right alongside the soft drinks—making it easy to swap one for the other. Consumers can be healthy with their stop-by-the-convenience-store-and-purchase-a-drink habit.

There have been swapping failures over the years. If you aren't old enough to remember carob, you aren't missing anything. It was supposed to replace chocolate. It didn't. The failure of carob (and the failure of some of our most misguided home-habit remedies, such as thinking we can just replace Fritos with carrots in our kids' lunches and they won't notice) is a lesson in how all of these habit-formation techniques must be *organized*. When we swap, we have to remember the reward principles from chapter 8. If a new option is noted to be a marked downgrade, dopamine neurons decrease in activity, signaling to avoid that action in the future. When we attempt to create a new cue for a response, we have to remember its bigger context. All of these pieces will help to establish the recurring consistency needed as the bedrock of habit-formation cues.

10

Total Control

If we are facing in the right direction, all we have to do is keep on walking.

—Joseph Goldstein

Mise en place is French for "put in place." This idea permeates professional kitchens. Chefs don't start cooking until everything is, literally, in place: their implements at the ready, ingredients measured and chopped, and items ordered as they are used in the recipe. *Mise en place* reduces friction in the kitchen. It removes the restraining forces that get in the way of making a recipe and sets up the driving forces to cue it automatically.

This is a deceptively simple concept. But beginning chefs don't naturally understand friction. Instead, the new students I met at the Culinary Institute of America at St. Helena, in Napa Valley, wanted to jump into a recipe and start creating great food. I talked with Robert Jörin, associate dean and professor of baking and pastry arts, about how beginning students work. "They're looking, okay, it's flour, it's sugar, probably the first ingredients of the recipe. So they go and get sugar and flour. Then they start mixing. Then they get down to, 'Oh, I was only supposed to take *half* the sugar and mix it.' By then, they have to start over." The ingredients are wasted and time is lost. "They don't look far enough ahead in the recipe, so they don't set it up right."

As a professional chef, Jörin says, "My first thought is *mise en place*: 'What do I need to make this?'" He prepares the kitchen for cooking so that it's easy to complete the recipe. "Once I know that I have all the ingredients and all the equipment to make a new dessert, then I mentally figure out in which order I need to do it. I have it scaled out in the way that logically I am going to use it. When I start working, I didn't forget anything. It is lined up in front of me so that I don't have to think about it. So, here is my crunchy layer on the bottom. Then this is my filling that goes on top of it, and then here is whatever glaze that goes over it." When cues at your station are organized, "you can concentrate on the methods that it takes to make the dessert rather than have to worry about if you have the right ingredients on the right tray."

Students learn this friction-reducing approach on the first day of class. Jennifer Purcell, the director of education at the Culinary Institute, explained, "We do mental repetition. Also, we physically repeat the competencies. All of the ingredients are within a very short reach, very tight. You don't want a lot of extraneous movement. You want to be able to work quickly, comfortably, with the least steps or exertion. A chef wants a flow of movement that becomes natural, comfortable, and almost without thought."

Professional kitchens run on a model of automaticity. They repeatedly and quickly turn out the same quality dishes to keep a restaurant full of customers happy. To do this, chefs harness the external forces in their kitchens by creating stable contexts that automatically cue the right response.

But this is a principle that has power beyond the kitchen.

Jörin explained that he uses *mise en place* in his job as a teacher. "Every day when I go home, I set up my roster, my whole class for tomorrow or for Monday. All the stuff that I need for Monday morning is on my desk ready to go. This is how I live my day. I want to know what I do at 10:00 tomorrow. I don't want at 9:00 to send out a text that this person has to be over there. In order to efficiently get stuff done, you need to have a timeline and set up for different tasks ahead of time."

Jörin says this is also how he ran his own bakery before he started teaching. "You can't run an operation if you are not ready. You can't wait until Monday morning to start. We have more transient kind of people [in this business] who tend to move around. I'd rather have everything set up so that, no matter who is going to be here on Monday morning, my customer's going to have food. So I have everything ready to get it done. That's what's instilled in you when you work in this industry. If you have five hundred people who are hungry, you better feed them, because they don't take no for an answer."

Harnessing friction offers a whole new way to think about changing behavior. The promise is that, by altering contexts that create friction in our lives, we can learn to automatically repeat rewarding actions. But first, we have to identify these contexts. And they are not always obvious.

If this sounds like a lot of work for your executive, conscious mind, you're absolutely right. Preparation in a kitchen requires calling on that part of ourselves that projects forward, that plans, that sees patterns, anticipates failures, addresses weaknesses, and designs stopgaps. The initial starting position for some of your most successful habit formations will be highly rational and will call on your conscious self.

The benefit of the habitual self is that it builds on that starting position and eventually obviates a need for continuing attention. A lot of up-front investment pays off with passive returns for all time.

*

Sometimes information can look like friction. But as we saw with the 5 A Day program for fruits and vegetables, it is not the same. Doing is different from knowing.

A common recommendation to save money, for example, is to avoid using credit cards. After all, credit was designed to reduce friction on spending money, so that consumers can continue to do so even when they have no funds. So savers are advised to use cash instead.

But what about debit cards? In some ways, they are similar to cash. Whether you use cash or debit, the amount of money you have available is immediately reduced, and you have less to spend in the future. So the two are essentially equivalent. But they differ in how easy or difficult it is to actually make a purchase—in the friction they provide. Students in one study were willing to pay about 30 percent *less* for coffee and beer when using cash instead of debit cards.[1] It was as if they valued the items less and thought they were worth less money when they had to hand over bills. So they were not willing to spend as much for them.

What is it about cash that puts friction on purchases? For one thing, we have visibly less of it in our hands after buying something. Using plastic does not have such tangible effects. Use it repeatedly, and it looks the same. Also, when we go to make a purchase with cash, we have to decide whether to use large or small bills and maybe search for change. All of this is friction on buying. Turns out that the advice to switch to cash actually works. When we have to hand it over, we aren't willing to pay as much for an item. Having only cash on hand becomes a driving force to save money.

Other advice we get is not as effective because it doesn't necessarily change how we do something. Adding calorie counts on menus should logically encourage us to eat fewer calories. New York City provides a test because, since 2008, calories have to be displayed on chain restaurant menus. A survey of more than seven thousand fast-food customers in the city showed that when the regulation was initially implemented, the information was noticed by 51 percent.[2] That declined to 37 percent by 2014.

Regardless of what customers noticed, labeling had *no effect* on their behavior. When purchases were compared over six years at restaurants with and without the labels, customers at all locations actually increased the calories they purchased at each meal. Calorie counts also did not decrease how often people ate out each week.

Of course, information can influence us when making occasional, large purchases. That yellow EnergyGuide sticker on your refrigerator or washing machine? It gives all kinds of helpful information about electricity use and operating costs. For big purchases, we consciously make decisions about one model or another. But even here, the effects aren't as large as we might want. Consumers have to weigh the abstract information about future energy use and savings on, say, a refrigerator against immediately compelling features like the price tag, the color, and the presence of an ice maker. Nevertheless, energy and water efficiency labels move consumers somewhat toward purchasing more efficient products.[3]

Regardless of their influence on consumers, calorie labels on food and energy ratings on appliances are not wasted in the marketplace. Even when consumers show limited interest, producers show more. This information is a kind of accountability: Calorie ratings are an acknowledgment of healthfulness. Energy ratings are a disclosure of efficiency. After providing calorie labels, some chain restaurants altered their servings, so that we now see some smaller pastries on Starbucks' shelves.[4] With energy ratings, appliance manufacturers started to make more efficient products.[5]

Call it trickle-down habits. The corporations changed theirs, and in so doing invisibly changed our environments. The end result is that you and I changed our own purchasing habits.

✻

Mise en place works for chefs, but is it possible for you and me to control the friction on our own behaviors? Angela Duckworth and her fellow researchers asked a group of University of Pennsylvania undergraduates to list academic goals, such as "study French for an hour every night" or "finish all homework the day before it is due."[6] For a week, some of these students were instructed to modify their study spaces to minimize temptation in order to meet their goal. These students changed the external forces in their contexts by setting reminders or alarms, installing online apps to block distractions like Facebook, or perhaps reserving study carrels in the library. They set driving forces or removed restraining ones. A second group of students was told to rely solely on willpower and their ability to resist temptation. This is, of course, the way most of us spontaneously try to get work done.

At the end of the week, students rated on a scale from 1 (extremely poorly) to 5 (extremely well) how successful they were at meeting their study goal that week. On average, all students were reasonably successful, but those controlling their situations scored about half a scale point better than those who tried to simply buckle down with self-control.

This *situational self-control*[7] seems like an indirect approach—modifying the world around us instead of what really matters, our own behavior. Just as with the beginning chefs (and my cousin on Facebook), our intuition is to jump in immediately and act on a new resolve. Students in the research project above were similarly inclined.[8] When high school students explained how they handled a recent self-control challenge in their lives (mostly interpersonal conflicts or academic problems), their most common responses involved changing themselves: 38 percent said they tried to change the way they thought, perhaps motivating themselves by laying out the pros and cons of doing homework. Twenty-four percent said they tried to change their actions, perhaps exerting self-control to restrain themselves from retaliating against another student who had upset them. Only 16 percent said they tried to change something about the situation, and only 12 percent tried to find a new situation.

Maybe you want to have a happier relationship with your spouse or partner? If you rely on motivation and control to do this, you will restrain your impulse to make a critical comment when he or she does something irritating and instead try to express warmth and appreciation to him or her. Or maybe you want to stop procrastinating at work? Relying on the same approach, you would just inhibit the impulse to check social media or to hang out with your overly talkative coworker. We set clear goals and then *effortfully* control our actions in order to attain them.

But behavior change through self-control, as University of Pennsylvania students experienced, isn't as successful as behavior change through altering contexts. Even if it were equally effective (which it's not), controlling our actions simply isn't fun. It means we have to continually fight our desires. It means we have to be eternally vigilant, wretchedly stopping ourselves from doing what keeps coming to mind. It means we have to be wet blankets on our own enjoyment.

In the study, Penn students who changed the spaces where they studied were not in this unhappy state of war with themselves. After adjusting their physical and social surroundings to remove temptations to play instead of study, students said they didn't experience many unwanted desires. They were not, for example, torn between watching a movie with friends or studying for a test. They had put themselves in the library, where no such tempting possibility existed. They didn't have to consciously force themselves to do the right thing. Instead, they did what was easiest in that environment—study. They didn't have to conquer themselves and deny their urges. They didn't have to be wet blankets, because they had no fire to put out.

For twelve years, I drove a Honda Civic hybrid, one of the first such models. I was proud of that car and loath to give it up. My husband finally convinced me that I needed a car with more safety features. My new car beeps a warning whenever I get too close to an obstacle. Turns out, there's friction in collision detection.

The beeping was annoying at first. I complained a lot about that car, especially to my husband. But eventually I got used to it, and I don't even hear it anymore. The last time I rented a car, it did not have the warning system. I didn't notice its absence until I was backing out of a parking space and hit a brick wall. Without the warning cues I had grown accustomed to, I ended up with a sizable dent in the bumper. That irritating beep alert provided useful friction that, when no longer there, left me with quite a repair bill.

Once in place, the forces in our environment continue to cue us to achieve our goals. We can ignore them or take them for granted, but they are still automating our behavior long after we have forgotten them. And yet, many of us discount the important role that such forces play in our behavior. Instead, we remain in the trenches, fighting to stay motivated and to exert control.

*

In chapter 5, we discussed people who report having a high level of "self-control." They are especially effective at achieving health, wealth, and happiness. Their lives are marked by success on many fronts. We found that these individuals *don't* attain these admirable outcomes in the expected way—by actively exerting willpower. Their success is not due to some superhuman ability to resist urges and inhibit unwanted actions. As we learned, people who score high on "self-control" scales aren't using control at all. This is a misnomer. Instead, they form habits to automate their behavior. Habits make it easy to accomplish their goals.

There's an important sequel to this story of how people with high "self-control" are successful. It has to do with contexts. It seems that the talents of high "self-control" people extend beyond just knowing how to form beneficial habits. They also seem to understand how to put themselves into contexts with the right forces to achieve their goals.

In an online survey, individuals who scored high on a "self-control" scale also agreed with statements like "I choose friends who keep me on track to accomplishing my long-term goals," or "When I work or study, I deliberately seek out a place with no distractions," and "I avoid situations in which I might be tempted to act immorally."[9] These people understood the power of context cues to make actions easy or difficult. They recognized that if they controlled their surroundings, they'd control their actions, too. Once someone understands this, it gets easier to form beneficial habits. Students who tested low in "self-control" didn't agree as strongly with these statements. They weren't trying to make their lives easier by establishing the right external forces—ones to drive desired behaviors and put friction on undesired ones.

High "self-control" people don't just say the right thing. They do it. Students in one study could earn up to $25 for quickly solving a list of anagrams.[10] They had the option of starting work immediately in a noisy graduate student lounge or waiting five minutes for a quiet room to become available. Students who had scored higher on a "self-control" scale mostly decided to avoid the noisy lounge. They wanted to wait for the peaceful place where they could concentrate, even if it cost them time. Same with students taking an online intelligence test.[11] They were given a choice between a plain form or one decorated with swirling artwork. Again, those who scored higher in "self-control" were more likely to choose the plain IQ test. With the boring version, they could focus and perform their best. They made the right up-front choice for top performance, given that the distracting forms could only get in their way.

As you set out to develop new habits, you're going to quickly rediscover something that you intuitively knew in advance: the greatest source of friction in this world is other people. They are both helpful and detrimental forces on our desired selves. People with high "self-control" not only know this but act on it. College students in one study chose one of two partners to work with on a task (actually, experimental accomplices).[12] "Alex" was supposedly undecided on his major, spent his spare time playing video games and partying, and mostly slept late during winter break. "Taylor" was a premed student with a part-time job who volunteered at an animal shelter and studied during winter break. The guys seemed equally likable. But participants who had earlier scored high on "self-control" mostly wanted go-getter Taylor as a work partner, whereas low scorers were equally likely to select Alex the slacker or Taylor the achiever.

Not everyone recognizes the way our environments influence us. But like the University of Penn students, we can all start to benefit from this insight and gain the trained eye of someone high in "self-control."

*

If you leave this book with one word and one idea, I hope it's friction. It is simple and intuitive, and can be manipulated to help accomplish astounding things. The forces created by the contexts in which we live come closest to drawing in ideas from all parts of habit science. Their results are constantly on display.

In a study of diners at an all-you-can-eat Chinese buffet, about 42 percent of obese patrons sat facing the buffet, the food in full view.[13] Only 27 percent of normal-weight folks sat facing the buffet. Instead, thinner diners mostly sat with their backs or sides to the buffet. The thinner diners did other things that made it more difficult to respond to buffet cues: 38 percent sat in booths. If they returned for additional helpings at the buffet, then their companions would need to move. About half as many obese patrons (16 percent) sat in booths. Most chose chairs, making it easier to get up for food. Thinner people were also more likely to put napkins on their laps (50 percent), while only 24 percent of obese people did so. A napkin is a pretty minor impediment to getting up to return to the food. But as we've seen, even small adjustments to the environment can make a difference. The most striking difference was that 71 percent of normal-weight people browsed the entire buffet to see what was available before they dug in. This let them pick and choose what they wanted instead of just eating their way through everything available. Only a third of the obese people did this. Most started to serve themselves immediately, without checking out what was available first. They were less selective.

Controlling driving forces and imposing restraining ones is possible, it seems, even at an all-you-can-eat buffet. Although normal-weight diners could not actually remove cues, they were able to limit their exposure to them. In so doing, they did not have to make decisions but instead could eat as they might typically do, in normal settings.

The alternative, I guess, is to throw out all of the science and reality of habit formation and continue thinking that each of our destinies is ruled by our willpower alone. You could ignore the psychological forces in our environment and continue to believe that each of us is acting in a vacuum, with the only pressure coming from our own decision and will. So when you trip up and fall behind, you can feel terrible about yourself. And when you succeed, you can feel intrinsically superior to other people who are struggling. Does that sound good? More important: Does it sound familiar?

There's a much better way.

*

Habits lead to a better life. It's not just about productivity. You hear people often complaining that they overthink things. We all do, sometimes. It can lead to anxiety and become a real challenge to getting anything done. A new appreciation for "mindfulness" has cropped up in recent years as a kind of panacea for this "overthinking" menace. The idea is to be thoughtfully aware, not lost in your head. Being mindful is being focused on the here and now, and not ruminating on past mistakes or thinking forward to upcoming challenges.

Habits are perhaps the most natural and effective way humans have to achieve this nonevaluative state of mind. A habitual mind is a *benignly thoughtless* mind. It is a mind that sorts tasks into their proper places. It delegates. It sits at the intersection and assigns routes. It is not obsessed with figuring out when you fall asleep, as you might have tried to do as a child; instead it just responds to the sleep cues in your contexts, and you drift off as you usually do.

If your goal is to stop fighting so much with your partner, then you want to form a habit of quietly, mindfully listening instead of lashing out. You'll more easily form a listening habit if you don't deliberate about each disagreement, trying to figure out who's at fault and who should apologize. Overthinking makes it harder to be positive. And, more to the point, it also might stop habits from forming.

A study involving a children's video game about making sushi showed just how advantageous *not* thinking too much can be.[14] The game had sixteen steps—adding water, salt, and sugar, stirring, laying out the rice, and adding salmon. As players practiced, the avatar in the upper-left corner told them what to do. Jen Labrecque, Kristen Lee,

and I warned some players that at the end of the study, they would have to make sushi on their own. They needed to deliberately plan ahead and memorize the steps. Others got no instructions but blithely continued on and played the game ten times.

Participants who were instructed to memorize did not form habits as well as those who just repeated the game without overthinking it. We know this from a test of the strength of players' automatic cognitive associations. Participants chose the next step in the recipe as fast as they could after a prior step (vinegar, then sugar). Memorizers were slow to respond. They were apparently still thinking about the recipe, even after playing it ten times. Those who just practiced without much thought were significantly faster, suggesting that they were making choices automatically. For them, adding the salmon cued picking up the knife and cutting.

In more evidence that overthinking impeded habit formation, all players were asked to change up the recipe and add a new ingredient, either hot chili oil or soy sauce. They now had to change their behavior. For this part of the game, they were on their own. The avatar didn't tell them what to do. Out of three tries, players forgot the new ingredient almost 20 percent of the time. But not everyone tripped up in this way.

Players instructed to memorize were more successful at altering the recipe. Without strong, automatic cognitive associations, they just changed their behavior. Because they were trying to keep it all in their head, without using context cues to trigger the next steps, they did not form a habit that persisted. When trying to change behavior, we are tempted to act like these participants, plotting and planning our every move. It is as if we are trying to learn the tango by deliberating about each dance move. It just won't happen.

In contrast, players who just repeated the game during practice were more likely to slip up by forgetting the new ingredient. For them, the next step in the recipe seemed to just pop into mind (add sugar!), and participants acted on that before having a chance to think, "Darn, I meant to add the chili oil now!" They were driven by habit.

This research is in early days, and science has yet to reveal exactly how overthinking gets in the way of forming habits. But even rats form habits more readily when they don't have to attend closely to a behavior and determine whether it's the right thing to do to get a reward.[15]

The implications for habit formation are clear: habits are more likely to form when we act repeatedly without planning and deliberating.[16] Then we are able to relinquish control to the context, allowing our actions to be cued automatically. After establishing the right driving and restraining forces in the contexts of your life, for example, you mindlessly eat healthy foods, get work done on time, and show affection to your family. Overthinking is beneficial, of course, if you want to stay flexible and *not* form a habit. You can do the same thing repeatedly, but the thought shields you from habit formation.

*

Remember that goal of getting your whole family eating together— talking, sharing, relating? That's now your habitual reality. You established the four basic building blocks of habit by: (1) creating a stable context (one night a week, 6:30 p.m. sharp); (2) reducing friction (with you as a driving force; removing restraining forces by initially doing all the cooking and cleaning yourself); (3) making it rewarding (serving everybody's favorite foods on those nights; letting the kids invite friends if they want); and (4) repeating until it becomes automatic (even when the rest of the family was ready to mutiny against your bright idea).

The same principles apply to making the goal of taming your finances a habitual reality. To cut spending, you: (1) created a stable context (found a good cheap grocery store brand, cooked extra at dinner so leftovers were there for packing lunch); (2) increased friction (carried only cash); (3) made it rewarding (hosted movie rental nights with friends who share your love of indie films; felt pride at paying off your credit card balance); (4) repeated until it all became automatic. Then you took it a step further. You enrolled in your company's automatic 401(k) plan, you began to bring your favorite blend of coffee to work— you did any number of things that required an initial conscious decision, and then fell into the background of automaticity, saving you money as surely as earnings from a stable interest rate.

PART III

Special Cases, Big Opportunities, and the World Around Us

11

Jump Through Windows

Should you find yourself in a chronically leaking boat, energy devoted to changing vessels is likely to be more productive than energy devoted to patching leaks.
—Warren Buffett

For two days in the dreary late London winter of 2014, the Underground metro system shut down. The union representing its workers went on strike, and 171 stations of the 270 total were closed. The closures were not systematic or predictable. Some staff showed up despite the strike. But even though the shutdown was not comprehensive, it was still hugely disruptive. In a transit system, just one station closure can throw all kinds of customary routes into disarray.

Depending on your political persuasion, the strike was a huge success or a dismal failure. To us in habit science, it was unquestionably one thing: a fantastic natural habit change experiment.[1] Commuters the world over are extremely valuable real-world test subjects, because their desires are so uniform: They want a fast route to and from work. They want to not be commuting as quickly as possible. This is especially true for commuters making use of underground transit, which is typically more noisome, loud, and crowded than other forms. The London Underground is no exception. To make matters worse, unless you are an old hand at the Tube, the system is not intuitive to use. Maps aren't to scale. They show relative positions rather than absolute distances. Travel time is hard to estimate, because train speeds are variable. London is an ancient town, and spread out, and lacks the grid rationality of New York City.

And, of course, it rains. The first morning of the strike was true to form, and many commuters who had prepared for the strike with plans to bicycle or walk to work were forced underground by the weather. They had to figure out a new path to get to work, one that bypassed closed stations. A highly routinized part of their day was suddenly new again. What was once relegated to the habitual mind was now forced into the domain of the agentic mind.

Most commuters used a rechargeable transit pass called an Oyster card, which discounts trips for high-use travelers. Using the card data, researchers were able to track more than 18,000 regular morning commuters before, during, and after the strike. The disruption was monumental. On the days of the strike, only about 60 percent of these commuters were able to enter at their normal stations, and about 50 percent exited as usual. In between, commuters were improvising. Surprisingly, the collective improvisations didn't drastically increase commuting time. On average, people spent only 6 percent more time in transit. Some people actually got to work *faster*—especially the commuters who typically used slow lines or traveled on distorted areas of the map.

Of course, commuters could have experimented with alternative routes even without a strike. Only their habits were stopping them from trying different Tube lines or starting or stopping at different stations. But in the rush of daily life, we don't often take the time to experiment. We find something that works adequately, and we stick with it. For the sake of ease, we settle.

The closure of the Underground made this "adequate" way of doing things briefly impossible. This is called *habit discontinuity*—a term coined by researcher Bas Verplanken to describe how our habits are disrupted by changes in context.[2] When habitual cues disappear, we can no longer respond automatically. We have to make conscious decisions. We are open to change—even, sometimes, serendipitously finding improvement.

This chapter demonstrates how these discontinuities in our habits can, paradoxically, be the best possible thing for the development of effective habitual selves. They can disrupt our "just good enough" habits and make us seek a newer, faster, more effective way of doing things.

*

We don't need to rely on the vagaries of organized labor to experience our own discontinuity and renewal. Major life events—starting a new job, moving, getting married, having children—have the same effect, many times over. They take away our habit cues and remove the predictability of life. In chapter 10, we learned that when you wish to try something new, changing contexts is a good place to start. Without the familiar cues to guide us, we're forced to think and make fresh decisions. In practice, it can be hard to surgically remove certain cues from our lives, which is why these discontinuities are so valuable. They shake everything up, and for a moment, all of your behaviors—habitual and otherwise—are in the air, waiting for you to direct their placement.

Yes, major life changes are stressful times full of uncertainty. But they are also opportunities to reimagine ourselves and restructure our lives. We are freed up to practice new behaviors without interference from established cues and our habitual responses to them. Discontinuity forces us to think. By making fresh decisions, we act in new ways—ones that may work better for us.

Our lives already comprise many, many habits. Some we're aware of, and some we're not; some have outlived their usefulness but keep operating, often out of sight and out of mind. Big events in our lives are an opportunity to declutter our habit selves and free them up so that we can consciously establish some new, more productive habits.

Maybe you went out for drinks and dinner every Friday with friends from work. It was fun at first, something you looked forward to. But lately you've noticed that the conversation circles around the same few topics. You can't bear to hear, yet again, your friend's stories about her son, or the usual complaining about office politics. You've even started ordering the same things week after week, because you've tried everything else on the menu. What began as a welcome start to the weekend is now feeling like a duty.

Or perhaps you love watching the sunset over the lake by your house. You decide that it's a wonderful way to end each day. So you make a habit of sitting on the deck every evening to watch. But as time has worn on, that sunset has become slightly less enthralling. Eventually, your habit starts to feel constraining. Your partner has stopped joining you, and you have started to spend your time thinking of other things you could be doing instead. Sunset-watching has started to seem like an obligation. Even good habits can turn into ruts.

The little-known nineteenth-century French philosopher Félix Ravaisson was able to put this concept into concrete terms. He called it the *double law of habit*.[3] Basically it means this: repetition strengthens our tendency to act, but it also weakens our sensation of that act. In other words, we *habituate*. It's a deceptively complex process, and one that has power to sap force and meaning from our lives. We tend to keep doing things long after they have lost meaning for us. Yes, we can take advantage of that dynamic when we form new habits, as they lose their hard edges with repetition. But it's a double-edged sword.

Habituation is one reason we lose interest in the material stuff we buy (thinking those things will finally make us happy). Certainly, you enjoyed sitting on your new couch the day it was delivered. And you got to show it off to your friends the next time they visited. But after that? You probably don't notice it much now. It just got folded into the rest of your evening habits. It's literally part of the furniture of your life. You plop down on it to watch TV or surf the web.

Habituation also occurs in relationships. You regularly exchange greetings with people at work, pick your kids up from school and ask about their day, and perhaps even call or text your relatives at specific times. You establish *behavioral interdependencies* in which other people are a cue for your action and you in turn cue their response. "How was your weekend?" "Great, and yours?" or "How was school?" "Fine, Mom."

Over time, you come to think about these interactions less and less. You just do what you've always done.

Long-term marriages are marked by such stable interactions. As spouses repeatedly do the same things with each other, they start to think less about what they are doing. They get up together, eat together, and do chores for each other without putting in much thought. They don't have to wonder what their spouse will do. They just know from experience. Over time, their emotions start to wane as Ravaisson's double law takes hold.[4] Spouses are likely to find that they no longer feel the passion that marked the beginning of their relationship. As actions become more automated, couples have less need to think, and their emotion subsides.

What was regrettable but acceptable with regard to your new couch can become untenable in your marriage. It's not okay to simply settle into your spouse's presence.

In happy marriages, discontinuities can work some magic by reintroducing the romantic intimacy hidden by time. A brief physical separation is a temporary discontinuity. Maybe you travel for work or take a trip to visit parents. Brief conflicts or arguments also can represent discontinuities, assuming these are not unresolvable.[5] These changes spur partners to share their feelings and act in new ways. Spouses start to think about their loved one and relationship anew. This, in turn, makes them reflect on the basic motivation for their partnership—what led them into this arrangement in the first place. For most of us, that means love. Adding to this experience, couples often express additional affection for each other when reuniting or making up after a conflict—affection that is experienced all the more strongly because it is atypical. Those of us in happy relationships can benefit from this insight. We can create minor discontinuities with new experiences (sailing lessons? bridge? a reading group?) that spark us to do new things with our partners, share our feelings, and heighten our experience of romantic intimacy. Arguments might trigger the same dynamic, but why not skip the bad feelings and, instead, go to a cooking class together?

In unhappy marriages, however, discontinuities don't have such positive effects. Unhappy spouses habitually engage in destructive cycles that continue automatically even when they intend to do otherwise. Couples stuck in such unfulfilling relationships may recognize the damaging patterns but feel helpless to change. People also habituate to the emotions in these marriages and may not experience intense distress and heartache from seemingly toxic interactions. Perhaps you have watched couples who respond to each other with apparent spite and anger, all the while seeming to experience little of the emotion that goes with such interactions. They simply have become habituated over time. Discontinuity, such as a physical separation, brief conflict, or new experience, could spin such couples in multiple directions. It could free them up to address problematic relationship patterns, or it could lead them to split up for good.

Habit discontinuity gets us out of ruts by exposing us to the underlying reality of why we're doing what we're doing and why we're going where we're going. Life is a more intense experience once we're no longer on autopilot. But it's also less predictable. Our conscious self is now in charge, as we think, weigh options, and figure out how best to meet our current goals. Discontinuity removes old patterns in our lives and, by making us think, resynchronizes our habits with our goals and plans.

*

There's an old concept in economics called *creative destruction*. It captures the inevitable moments of stress and fracture created in a market economy. Those moments are immediately painful, especially to those directly affected. Stocks crash. Jobs are lost. Whole industries vanish. But from the privileged perspective of an observer, this destruction also contains the seeds of new growth. Innovation can look like failure—ask anyone in Silicon Valley. It's practically the ethos there.

Your habitual self is a crucible of this kind of destruction. After you see how, you'll be able to control the rate of both destruction and creation.

Once you get into the habit of driving to work, it's pretty automatic. You just get into your car and follow the route you always take. It requires effort to do something else. To take the bus, for example, you have to figure out bus schedules, fare costs, whether you need a special transit card, and how early to wake up in the morning. You don't face these decisions when you follow habit cues that keep you driving.

Enter the greatest discontinuity of all: moving. A study compared the transportation habits of 69 employees at a small English university, all of whom had moved in the prior year, to those of 364 established residents.[6] The researchers started by assessing the environmental values of all participants, and found that they held a variety of attitudes—some were quite eco-conscious, and others didn't care very much at all. The town had convenient transit options to get to work at the university, including a good bus system, along with cycling and walking paths. No one had to drive. Nonetheless, 60 percent of longtime residents drove to campus. Strong environmentalists as well as those who did not care all mostly drove their cars.

The recent movers were different—in a good way. Among those who said they were highly concerned about the environment, only 37 percent of trips to campus were in cars. These newcomers were more likely to take the bus, bike, or walk. When they had no habit to follow and were forced to make new decisions, their green values won out. Among recent movers who said they were *not* concerned about the environment, 73 percent drove. In the absence of habit, they were also true to their values. They did not try other, environmentally friendly forms of transit.

In new contexts, we choose behaviors that fit our current goals. We cannot easily repeat what we did in the past, and we have to more thoughtfully synchronize our actions with who we are right now. Another way of looking at this is that the discontinuity of moving actually made people own up to—and act on—their stated beliefs. Discontinuities can make us into more genuine, integrated versions of ourselves.

While we stand to benefit from discontinuities, most of us don't welcome them into our lives or seek them out. At best, you probably feel ambivalent about such life changes. And you should, given the double nature of discontinuity. While freeing us up to find efficient routes to work, experience anew love for our partners, and act true to our values, changes in everyday contexts are disruptive. Changes in contexts can leave us stranded, confused, and unsure of how to act. But there is value in understanding these effects—retailers and product manufacturers are already well aware of them.

*

For most of us, weekly grocery shopping is a study in efficiency. When 275 shoppers wore electronic devices to track their paths through a store, on average they covered only 37 percent of the total retail space.[7] Mostly, shoppers stick to the aisles where they want to make a purchase and bypass the rest. Grocery shopping is a chore, and we get out of the store as simply and quickly as possible.

But discontinuity happens when stores switch locations of items. Researchers evaluated what would happen when fruit changed location with vegetables, baked goods with cereal, and meat with salad mix.[8] Now shoppers would have to stop and think about what they wanted to purchase and where to find it. With the changes in store layout, shoppers would be introduced to new products they typically wouldn't see or purchase. They could no longer follow their automatic patterns. Researchers estimated that unplanned spending would increase by about 7 percent per shopper. Despite the already-fine-tuned layouts of grocery stores, disruption could make sales even higher. But changes in store design can also irritate shoppers, especially those over fifty, who are likely to lose patience when they can't find what they want.[9] It's a delicate game for retailers trying to get into your wallet.

Our shopping is also disrupted by changes in the design of product packaging. Radically new packaging makes it hard to identify an item that we regularly purchase. In 2009, for example, Tropicana changed the orange-with-a-straw picture on its Pure Premium orange juice. The new design had a glass of juice and more prominent display of "100% orange pure and natural." Consumers were surprisingly vociferous about their dislike of the change. They apparently started to think, "What does 'pure and natural' mean?" "Will this taste the same as my old OJ?" "Maybe I should try another brand." The fallout for Tropicana was reputed to be about a $30 million loss in revenue[10]—all because the brand decided to trumpet a putatively positive feature of its product.

Although disruption is more often than not a turnoff in the marketplace, everyone has felt the pull of a new must-have gizmo. We live in the iPhone era, after all, and our news cycle is regularly organized around a dramatic launch of some new groundbreaking piece of consumer tech. But this is actually well outside the norm of how new products enter our lives. New products are hard to launch. The lines of devoted customers that Apple is able to manifest with each new device is an astonishing success—one that flies in the face of how most of us encounter novelty in the marketplace.

New products create disruptions when we have to change our behavior in order to use them. In 2001, the Segway scooter was a *really new product* hyped by seasoned investors like Jeff Bezos of Amazon.[11] Steve Jobs predicted that cities of the future would be redesigned to accommodate its broad use. By 2004, however, only ten thousand units had been sold, and the Segway's destiny as a small niche product was clear. Compare this with the popularity of electric scooters, an *incrementally new product* that adapted kids' Razor scooters into a grown-up version with a motor. The scooter manufacturer Bird jumped in value from $300 million in March 2018 to $1 billion in May of that year and $2 billion by the end of June.[12] Other transport firms, such as Uber and Lyft, have introduced their own scooters. Of course, this difference in success could just be timing, given that electric scooters arrived fifteen years later. But research has shown that consumers have less favorable intentions to purchase really new products, and when they do say they will purchase them, they are less likely to follow through.[13] We just don't know what really new products will do for us, and this uncertainty makes us think and rethink our intentions to buy. As a result, we act in unpredictable ways.

*

There's a darker side to discontinuity. Researchers are finding that it can get in the way of some of the basic habits of good citizenship. In

Montevideo, Uruguay, people receive about three to six tax bills a year, for property, vehicles, persons, and sewage. Each bill is typically paid in person at local tax kiosks. It's not an efficient system. The average tax account in 2014 was about six payments in arrears, and only about 70 percent of municipal tax bills were paid on time.

In 2004, the Montevideo government tried something novel to encourage more citizens to comply. Using Uruguay's national lottery, Montevideo awarded winners who had paid their taxes on time the previous year the prize of paying nothing in the current year. This was a natural experiment that allowed researchers to compare 3,174 accounts that won the prize between 2004 and 2014 with 3,189 similar accounts that had to continue paying.[14]

Winning a tax-free year should have been enough of a reward to encourage everyone to continue paying on time in the future out of gratitude or civic responsibility. But it didn't quite work out that way. In fact, lottery winners, relieved of taxes for one year, were *less* likely to pay their taxes in the following years! Apparently, the disruption in payments started these citizens thinking about their taxes—and about evading them. They found it hard to start paying taxes again after a year off. They had to remember where to go, how much to pay, and when payments were due. The effects were not large: winning the lottery yielded a four-percentage-point reduction in future tax payments. But this affected the best revenue-generating citizens—the taxpayers who actually paid on time. Showing that the disruption in behavior was key, lottery winners who had enrolled in payment plans through their banks were not affected by the tax vacation. They had been paying automatically before the win, and their bank payments simply started up again once the tax vacation ended. Also, the reduction was not evident with vehicle tax, the one tax for which even winning citizens had to keep paying a minimal amount.

Citizens, it seems, develop habits of interacting with the government bureaucracy that have important effects. "For policy makers," the researchers warned, "lack of attention to habit can lead to perverse consequences."[15]

After learning the study results, Montevideo switched from tax holidays to giving rebates for prompt payment. Fortunately, the discontinuity effects on previous winners dissipated over time. About two years after a win, good-citizen taxpayers were back to paying on a regular basis.

The challenges of discontinuity to good citizenship hit closer to home when it comes to U.S. elections. Bad weather discourages voting. People just look out the window and decide to stay put. Rural and poor counties are especially affected. In nonurban areas, voters have to travel farther, and less wealthy citizens don't necessarily have access to transportation to keep them dry.

We can compare voter turnout during presidential elections in counties where it rained with sunnier counties. Even a millimeter of rain reduced voting by a little, 0.05 percent, in analyses from 1952 to 2012.[16] And this disruption in one year affected later voting. When people stayed home due to rain on one presidential election day, they were less likely to vote in the next one.

This is the dual nature of habit discontinuity. Disruption of cues in our everyday contexts can be beneficial, freeing us to act in more authentic ways. But disruption can also be harmful, wreaking havoc on our habits of citizenship, making tax evaders out of taxpayers, and increasing nonvoters in the electorate. There are gains as well as losses to be had when life events remove the foundational cues that control automaticity in daily life. These dual effects reflect a basic fact about habits—they have no essential nature as good or bad. Just as our habits run the spectrum from beneficial to harmful, so too does disrupting them, with polarities of course reversed. Discontinuity, however, is not just about your habit self. There's also a role to be played by executive control and your more agentic self.

*

Major life changes often sneak up on us. But we have some control, at least in how we respond to the change. Once we understand how discontinuity works, we can use these same dynamics selectively, protecting our valued, beneficial habits and altering unwanted ones.

Protection comes in multiple forms, as suggested by a study of students transferring to Texas A&M University from other schools.[17] Leona Tam, Melissa Witt, and I contacted transfer students one month before and one month after their move in order to assess the fate of their everyday habits, including habits of exercising and watching TV. Some students had strong habits of doing these things when we contacted them before they moved. Two months later, most of these students reported that, with the discontinuity of the move, they were no longer regular exercisers or TV watchers. But not everyone lost their habit. For some students, the specific context in which they exercised or watched TV was the same across locations. For exercise habits, they might have continued to work out in a gym or run on a track. For TV habits, they might have continued to watch on a screen in their bedrooms. When cues remained stable like this, so too did habits. Although we couldn't tell if students deliberately selected new contexts to be the same as old or accidentally stumbled into similar circumstances, the outcome was clear: with stable cues, habits were protected.

Not all habits are worth preserving. Exercise is one that most of us want to hold on to, but TV watching is hardly beneficial for students. The end result was the same for both: change in performance contexts disrupted habits, and stability preserved them, regardless of one's being healthy and the other a time-waster. This should be a familiar story by now—the habit mechanism does not discriminate between actions that are beneficial for us and ones that are harmful.

Our transfer students illustrated another way to protect habits—a way familiar to your executive self: through deliberately carrying out their intentions. Even without the familiar cues from their old college, students could still decide to exercise or watch TV. In new performance contexts, some students put into action their intentions. They were back to doing the heavy lifting, and potentially forming a new habit in their new home.

By understanding cues, we can keep valued habits in place even when disruptions occur in our larger lives. But sometimes we are looking to change. We can bring disruption upon ourselves by altering the contexts of our lives. And we do! Each year, about 11 percent of Americans move,[18] meaning that most of us live in one place for about eleven years.[19] We change jobs even more often, on average switching employment every four years.[20] Any such major upheaval provides a window of opportunity to undo bad habits and let some needed light and air into ones that have become stale. When we most want change, discontinuity is a friend. We might wish to stop smoking, quit our jobs and start new careers, or leave abusive relationships. It's here that we can take advantage of disruption. As a corollary to our new ability to preserve and protect good habits in trying times, we can also use those times to break up and tear apart old, unwanted habits.

Can you think of a time like this when you successfully made a sudden and dramatic change in your life? How did you do it? Was it pure decisiveness and determination, or did a change in contexts help out?

Or was there ever a time when you failed to make a much-needed transformation in your life? Did you lose your nerve or find it too difficult and complicated to change all the things that needed changing?

This is what researchers asked 119 adults from the Harvard Extension School.[21] Participants described many different changes, including in their careers, education, relationships, and health.

When people told stories of successful life changes, more than a third mentioned changes in context: 36 percent of the successful stories involved picking up and moving house, even if only for a few months. One person, explaining the timing of an attempt to quit smoking that succeeded, said, "I felt I would find it easier to quit smoking in a new environment which didn't have the usual cues and associations." Another made a move because "I loathed law school. I was physically ill for a good part of my first semester—all, I believe, stress-related. I was depressed as well. I found few friends in the law school environment—it was too competitive and cold for real friendship." An additional 13 percent reported making other changes to life contexts, such as finding a new group of friends or a different job.

The stories of unsuccessful change were very different. Only 13 percent noted moving to a new location, and none involved altering the immediate environment. Mostly, participants provided reasons why they could not alter their current situations. As one said, "Leaving my job, with the economy in the state it's in, seems rather risky because I have a lease and bills to pay." According to another, "It has been easier to fall back on an old job than to suffer the rejection of a job search and the confusion and difficulty of deciding on one field." Stories of failed change often involved feeling stuck in one's current environment. Fully 64 percent of the failed changers noted external circumstances that made change impossible.

These personal narratives of change and failure to change provide striking insight into the power of contexts. People successful at changing their behavior leveraged the opportunity of habit discontinuity. They changed contexts by going away for the summer, leaving their jobs, or moving. By eliminating habit cues, they gave themselves the freedom to make new decisions.

Yet personal accounts like these depend on people's recollections, and recollections are subject to personal mythmaking. We all tend to organize our life histories into story lines more intelligible than they really were at the time. For a researcher, objective data are more reliable. Luckily, there is one domain where that is possible—where hard data on the benefits and challenges of changing context cues are there in the numbers.

Major league baseball loves statistics. Because of that, the sport is a useful laboratory for measuring the effects of habit disruption due to a common occurrence—player trades. Changing teams disrupts a whole range of habit cues, involving teammates, playing fields, coaches, owners, fans, and residences.

To test whether trades also alter players' performance, researchers analyzed the records of 422 major league players from 2004 to 2015 who had season-over-season declining performance before changing teams.[22] These were top athletes in need of a change.

Before and after the change, researchers assessed batting average, the players' ability to reach base, and overall offense relative to other players. Men with declining performance who changed teams enjoyed significant improvements in all three indicators. For example, batting averages increased over two years from a low of .242 to .257. (For context, Mike Trout, one of the highest-paid players in baseball, with a salary of $34 million, bats an average of .312.) In contrast, a comparison group of 922 players with similarly declining records who stayed with their teams showed significantly smaller improvement.

For some of these players, the change was their decision. They were free agents and chose to move on. Others were traded away. Habit discontinuity worked regardless of the reason for change. The new cues were followed by a performance boost.

Again, there was symmetry in disruption, affecting good and bad habits alike. In a second part of the study, researchers tracked 290 major league players with season-over-season stable or improved performance. For these men, changing teams didn't help. In fact, it led to a drop in batting average and the other offensive metrics.[23] For example, averages dropped over two years from a high of .276 to .263. This decrease was much larger than that experienced by a comparison group of 1,103 players with similar past records who stayed with their teams. Again, it didn't matter whether the players who changed teams were free agents to decide or got traded. Good performance was disrupted by context change. Players got worse. The grass was, in actuality, not greener for these guys. For already-successful players, a change of scenery hurt.

Freedom from the confines of unproductive contexts turned around habit-based failures among professional ballplayers, who are highly trained, achievement-oriented individuals. It makes sense that they were able to benefit from a new team environment. But habit discontinuity can also devastate habit-based successes. Even professional athletes are susceptible. Athletes with rising performance who joined a new team were set back.

The lesson from all of this is that habit discontinuity is powerful. It alters the balance of habit and decision-making in our lives. Disruption makes us think. In so doing, it can make life interesting and allow us to act in ways that more closely reflect our values and interests. But it can also put beneficial habits at risk. Disrupting a habit is, of course, just the first step in making a change. It clears the decks and puts old habits behind us. How well we use that opportunity depends on what we do next. By understanding disruption, you will be able to (1) protect your good habits so that they can weather change, and (2) use disruptions to pierce your bad habits at their most vulnerable places.

The discontinuities in this chapter are often unwelcome in our lives. Losing a job or moving can be an enormous challenge to our stability. By approaching these changes from a habit perspective, we can see that they are nevertheless also excellent opportunities to remake ourselves, to literally become the person we've been wanting to become. We're more pliable, and our habit selves are more commandable. The destruction to our status quo is very real—the creation in its wake is entirely up to you.

There's one more piece to this, which tips the balance of disruption's results toward positivity. Have you ever had an internet outage at your home for a few days or even a few hours? Or have you showed up at that old beach house that your friend invited you to, only to realize when you get there that the Wi-Fi router is from 1997 and has all the signal range of a toaster? Avoiding the obvious missteps (*The Wi-Fi is down? Better mix a martini!*), you now realize that those precious few moments of new behavior in the aftermath of a disruption can be the first brick in a new pathway. It's a chance to improvise solutions that you want to stick to in the future.

Perhaps you pick up that ancient copy of *Moby-Dick* that someone left by the couch many summers ago. You start reading. After a few pages, your annoyance fades away. You realize, with some guilt, that this is the first time you've picked up a classic in years. You are on your way to starting a new reading habit—one that you could have started all along, but you needed the disruption to help you realize again how much you enjoy simply reading a good novel.

12

The Special Resilience of Habit

The storm is a good opportunity for the pine and the cypress to show their strength.

—Ho Chi Minh

Life is stressful. It never seems to go according to plan. It just doesn't unfold as we expect. Our preferences go unheeded—except when they seem to be randomly fulfilled by happenstance. The passage of time and the events it orders don't flow through any predictable course or canyon.

We now have diagnostic tools to quantify our experience, and these tools consistently tell us that, yes, life is stressful. In a recent survey, about 25 percent of Americans reported extreme stress.[1] And most of us say we experience more stress than is healthy. The causes are predictable. In 2017, more than 60 percent of Americans mentioned being stressed by the future of the nation, money problems, and work difficulties. Japan even has a word, *karoshi*, for the extreme workplace stress that leads to death. People are reporting more symptoms of stress than in previous years, including anger, anxiety, and fatigue. That last feature is not just a state of mind: our bodies react to stress with a stream of hormones including adrenaline and cortisol that affect our thoughts, feelings, and actions. Stress degrades our executive selves, or those higher cognitive processes involved when we plan, think ahead, and flexibly act to achieve our goals.[2] Our decision-making suffers.

Over the past few years, most of us have become aware of the health impacts of stress. Its harm has been widely acknowledged but seldom solved. Sure, there are retreats to go on and mind-sets to adopt, but these work only under certain conditions and for certain people. More to the point, many of us don't have the resources to go to a meditation retreat in a leafy New England spiritual center.

Wouldn't it be helpful if all of us carried around the tools to build our own anti-stress refuge, shielded from the slings and arrows of daily chaos? Wouldn't that be a great place to put the behaviors you want to keep up through good days and bad—the kind of behaviors that achieve long-term goals?

Actually, you already have that. Habits are those safe harbors in stressful times. They aren't affected by stress like our more conscious selves. In fact, they thrive. There's even a *boost* in habit performance when the rest of our mind is drained by life.[3] It's a special quality that makes habits particularly well suited to the daily grind of being our best selves. It's easy to imagine how a habit boost was evolutionarily adaptive for human ancestors (see bear—quickly hurl spear).

Just as *habit discontinuities* disrupt the cues to habit performance (chapter 11), so too does stress disrupt our conscious selves. It shifts the balance of habit and conscious thought. Each system, it seems, thrives under slightly different conditions. Under stress, habits stay online despite the fact that consciousness falters. For researchers, this pattern is an intriguing sign of the dissociation between habits and deliberation. For everyone else, this backup-system arrangement has clear practical benefits. With a habit, you are never left without a response, even when stress, distraction, or mental tiredness is derailing your conscious mind.

To study the intersection of stress and habit, one group of researchers had college students immerse their hands up to the wrist in ice water for three minutes or as long as they could stand it.[4] This was physically stressful, as you can imagine. To add social stress, students were videotaped and watched by an unfamiliar person while in icy discomfort. For contrast, a control group of students wasn't subjected to either of those stresses. Instead, their hands were lowered into comfortably warm water.

In the next part of the study, everyone practiced a computer task in which they chose certain shapes on the screen by pushing buttons. There was a reward: when a student chose the correct shape, he or she got a squirt of orange juice or chocolate milk through a straw within reach of their mouth. A bit unusual, but exactly the kind of immediate reward that easily forms habits. When they chose an incorrect shape, students got a less desirable drink of bland peppermint tea or nothing at all. With this simple task, everyone learned to choose the rewarded shapes. The earlier stress didn't get in the way of habit learning.

After students made fifty choices, the task changed, and rewards stopped. It no longer mattered what choice they made. Students who had not been stressed got the hang of this after about five tries. Yes, they acted on habit initially, but after just a few reward-less choices, they started to notice and alter their behavior. Participants began thinking that maybe if they chose another shape, the rewards would start up again. So they were prepared for the next trial. They stopped responding habitually and started to explore. They deliberately chose different shapes in the hopes of finding a new one that would give them a treat. In short, they were adapting to their new conditions, and feeling out how to get back to a rewarding experience. But the stressed students simply persisted with the same old habit. Their conscious minds were still focused on the painful experience and the assault to their system. They weren't able to flexibly consider alternatives.

Real-world stress yields similar effects. In a study of 174 tough decisions made by corporate executives about acquisitions, major product launches, or restructurings, the executives who felt more anxious and under the gun (based on spouse interviews and company reports) were less likely to take strategic risks.[5] In business jargon, anxious executives continued to *exploit* what had made the company successful in the first place and avoided *exploring* new innovations and growth.[6]

That kind of exploit-over-explore outlook is likely to leave the company larder bare of new products and put it at risk of becoming the next Blockbuster, Polaroid, or Compaq.

Stress has these effects because it influences what parts of the brain are active. Under stress, neural activation shifts away from regions involved in decision-making and goal pursuit (orbitofrontal cortex, medial prefrontal cortex, hippocampus).[7] Activation increases, however, in the striatal neural systems involved in habit responding and rewards. This combination tilts us toward a kind of autopilot. Our decision-making systems narrow to what worked in the past. With a stressor present, our minds are going to want to deal with it by shutting it down or moving far away from it. We become preoccupied with defending ourselves from the stressor and pay less attention to what is going on around us.

Unfortunately, in the modern world, these stressors are often situations that require quick thinking and complex thoughts. With a family member in the hospital, for example, you need to make decisions quickly. Or maybe you recently got laid off and, with bills looming, have to find another job immediately. Perhaps your stress comes from your partner's unhappiness and a potentially broken relationship. These place increased demands on your conscious decision-making. The threatening situation takes up your attention, keeping you focused on replaying or repressing the experience and able to think about little else.

Like the students in the experiment, your hands are in the ice water.[8] You have to figure out how to cope with stressful experiences. Habits can help. In the next part of that study, after making the ten choices with no rewards, students started to get rewards again for choosing the right shapes. The non-stressed students quickly learned about this additional change and switched from exploring novel shapes back to exploiting the habit they had learned earlier. Their adaptability got them around to the right strategy again, after a period of experimentation. The stressed students, however, never made the shift away from the winning strategy in the first place. They just continued to repeat the habit, which now got them the reward.

Perhaps we celebrate the imagination and initiative of the non-stressed students. We all hope to have the presence of mind to adapt to our surroundings and seek out new strategies. But we do not live stress-free lives. From a habit perspective, the more important insight comes from the other group of students. Through several disruptions, through stress, through rewards and no rewards, their established habit persevered. It didn't falter when their mind was surely occupied with their discomfort and embarrassment. Habit was resilient. It lasted through thick and thin.

Now imagine your own ice-water situation: your health scare, your job setback, your relationship troubles. Instead of selecting correct shapes in some laboratory, imagine you've established exactly the kind of healthful habit that carries on the necessary work of keeping your life together as you deal with the complexity of your source of stress. It's exactly the kind of hard work that your second self can quietly do, and as we've seen in this chapter, that work can be done even when the more conscious reaches of your mind are under stress. That's very good news for us. It should make you optimistic about the next time you go through a challenging period. You'll know that your habits and those parts of you that you've set toward your long-term goals will continue. Your beneficial habits will keep grinding forward, ignoring the drama of the day. Habit then becomes more than the robust fallback system that allows us to keep acting despite the challenges that life throws at us. It's the desired choice of both of our selves.

*

A few years ago, I had a neighbor who was a professional cyclist. She was really fast. We used to take rides together on days when she was resting and was keeping her heart rate low. With me, she would take her touring bike, not one of her racing ones.

At the beginning of a ride, we would find an easy pace, and it was fun. We told each other stories about our families. That lasted until we were about an hour into the ride. When we started to head back home, she always sped up. She would soon be way ahead of me, and no longer able to ride and talk. She was back to racing speed. When

I asked her why, she explained that on the first part of the ride, she consciously tried to slow down. After all, her rest day was an important part of her training. But as our ride progressed, the conscious effort to stay at my pace became too much. Her legs automatically sped up. She was simply too tired mentally to maintain my speed any longer. The irony is that she was now working harder physically, but as a habit, it seemed easier to her.

When we're tired and stressed, we expect to relapse into bad habits. We've all had the experience of acting this way. Being late for an appointment, we repeatedly press the elevator button as if that could make it arrive sooner. Hurrying to our destination, we keep pushing the walk button to turn the traffic light green so that we can cross the street. Frustrated in a traffic jam, we honk our horn over and over despite knowing that everyone is equally stuck. Under the gun, we act out of habit, whether the behavior is something that brings benefit, brings harm, or has no effect. The habit mechanism does not discriminate between responses that are likely beneficial in the current situation and ones that aren't.

In a test of how stress and weariness exploit both good and bad habits, students at UCLA's business school reported on their morning routines.[9] For seven weeks, students reported what they ate for breakfast and what sections of the newspaper they read before going to class. Two of these weeks were especially draining because students took multiple exams.

Exam weeks boosted reliance on habits. Students with strong habits for eating particular healthful foods for breakfast, such as hot or cold cereal and health bars, were more likely to do so during exams. Those with habits of eating unhealthful foods such as pastries, pancakes, or French toast, and drinking coffee with sugar, did the same. Paper-reading habits were exploited in just the same way. Students with the habit of reading an educational section of the newspaper, such as world news, were more likely to read their habitual section during exam weeks, as were those with the habit of reading entertaining, less educational sections, such as advice columns. Students without strong breakfast or newspaper habits did not show this boost during exam weeks.

The increase in reading is surprising. During exam weeks, students presumably were studying more and had less time for reading the paper. Nonetheless, students were *more* likely to read what they habitually read. This makes sense if you think about how stress affects habit. During exam weeks, students were less able to make conscious decisions about what to read. Students who typically read business news, for example, were less likely to remember to check on a particular local story of interest. As a result, they deviated less often from their typical reading matter. They just woke up and read the business section as they always did, probably while ruminating about studying and the impending exams.

More direct evidence of the habit boost comes from a study in which Duke University students identified four desired behaviors that they were trying to perform to meet an important goal and four unwanted behaviors they were trying to avoid.[10] For example, starting homework right after dinner was a desired behavior to get good grades, whereas playing video games was an unwanted one. Students also rated the habit strength of each behavior by noting how often they had done it in the past in the same place. The study lasted four days. At the end of each day, students reported (yes/no) whether they had performed each of the behaviors they had listed.

On two of the days of the study, students' cognitive resources were drained. They were instructed to use their nondominant hand to do simple actions like calling on their cell phone, moving a computer mouse, and opening doors. This was mentally tiring because students had to inhibit the impulse to use their dominant hand and instead remember to use their other hand. To make sure that they followed instructions, participants signed a contract and created reminders for themselves.

On the two days when students used their nondominant hand, they performed more habitual behaviors—both desired ones that met a goal and unwanted ones that interfered—than on the other two days of the study. Students who were tired from continuous efforts to use their nondominant hand were plagued by bad habits but also benefited from good ones. Mental tiredness, much like stress, boosted habit performance, reflecting the limited capacity of conscious thought and the hardiness of automaticity.

✳

Habit resilience sounds pretty good, but in some ways it's a new spin on an old, unfortunate tendency that most of us have, one that is exacerbated by the proliferation of distractions available to all of us these days. One "ping" from your phone, and you're following a notification to a new social connection and beyond.

Distraction, in habit terms, is a surfacing of the habitual self at moments when we would, were we in command of ourselves, rather that it stayed in the background. No one is safe from it (because almost no one is so powerful as to be in complete command of their habits). For most of us, distraction is a low-level life irritant. However, some people naturally live their lives in a sort of absentminded abyss, with their decisions often derailed by distraction. There is even a scale to measure this tendency.[11] You can check out your absentmindedness score at www.ocf.berkeley.edu/~jfkihlstrom/ConsciousnessWeb/Meditation/CFQ.htm. If you answer "very often" to many of the items, then you are likely to be the kind of person who is chronically thinking about something other than what you are doing.

In our daily lives, distractions are mostly just inconvenient. We climb into the car intending to go to the store, and with the ping of a text, our attention is diverted—and we automatically steer toward the highway that takes us to work. Or we walk into a room to get some item when a song we love comes on the radio—and we unthinkingly pick up the wrong thing. When people kept records of such slipups, they reported about one per day.[12] People who scored high in absentmindedness, however, had many such slips.

In some situations these days, distraction becomes something graver. During a visit to the clinic or hospital, your doctor likely has one eye on you and one on the computer. Most clinics now require electronic records. And these are beneficial, providing a continuous health care story of your life. But completing forms takes your doctor's attention at times when you need it the most.

Distracted doctoring extends beyond your medical records. Faculty and interns from a prestigious teaching hospital were surveyed about their cell phone use during medical rounds.[13] Nineteen percent of medical residents and 12 percent of attending physicians believed they had missed important information about patients because of distraction from smartphones.

Under these conditions, surgery can be a real risk. In a survey of medical technicians, about half admitted to talking on a cell phone during heart surgery when they were supposed to be monitoring bypass machines.[14] Similar numbers reported texting. This, despite 78 percent acknowledging that such phone use was dangerous.

Yet some of us end up in the hospital in the first place because of our own distractions. Hospital admissions from pedestrians' cell-phone-related injuries tripled between 2004 and 2010.[15] Intake records tell the sad stories: "28-year-old male walked into pole talking on phone and lacerated brow"; "14-year-old male walking down road talking on cell phone, fell 6–8 ft. off bridge into ditch with rocks and water, landed on chest/shoulder, chest wall contusion"; "23-year-old male walking on the middle line of the road talking on a cell phone and was struck by a car, contusion hip."

Consciousness distracted by technology leaves us acting on habit. Often that habit is as simple and mechanical as *keep walking forward*. That works well for us when the path is flat and clear of obstacles. But when terrain changes and a mindful decision is required, there's potential for a serious accident, or at least embarrassment.

A 1984 study analyzed letters from sixty-seven people claiming that they had been wrongly accused of shoplifting.[16] Many argued that they had inadvertently put items in their pockets and bags without intending to steal. More than half blamed the incident on distractions. Of course, there were no smartphones then, but several said they had just lost their child in the store. One had knocked over a display. One reported seeing her ex-husband in the store with another woman. Such events could leave people responding to habit cues without thinking, so that they leave stores forgetting paid-for goods or change; leave home unprepared without money or credit cards; automatically grab someone else's shopping cart; take a look-alike item from the shelf that was not the one intended; and perhaps even shop without paying. Distraction is at the heart of all of these, and habit resilience is at the heart of distraction effects.[17]

Distraction paired with strong habits mostly works fine. After all, we usually get home with our purchases and pocketbooks intact. But habit can do only what it has done before. New, look-alike packages ("Those dog treats looked like dishwasher pods!") cue our purchase. We end up at home with something other than what we wanted. Or we miss opportunities, perhaps automatically choosing our standard items without realizing that other, perhaps preferred items are on sale this week.

Online, distraction can be even more troublesome. We all get phishing emails, for example. They seem legitimate but ask for sensitive information or insert malware into our systems when we hit an innocuous-seeming hyperlink.

Subject: Please verify your account

Dear Student,

There is a technical issue concerning your university email account, which requires your attention. Please click the link below to reset your account and address the issue within the next two days.

http://mxni.nm/90SJOjk

Thank you.

A class of students at the University of Buffalo, in one study, was sent individual phishing messages similar to the one above.[18] Fully 83 percent of the students clicked on the link. Students with stronger email habits, who reported using email often and automatically, were more likely to click on the link. Students were especially susceptible when they reported paying little attention to the email and quickly deciding whether to respond. When our conscious decision-making is distracted, our email habits can be exploited by others.

Social media habits also make us vulnerable. A class of college students received phishing messages on their Facebook accounts.[19] First, each got a "friend" request. Two weeks later, they got a request from the same account for personal information in the guise of an internship possibility: "If you are interested to intern and would like more details, please reply with your student ID number, e-mail user name, date of birth within the next three days." Habitual Facebook users, defined as ones who used the site often and typically at the same time of day, were most likely to comply with both requests, ultimately sending personal information to someone they did not know. These same students said they were concerned about privacy on Facebook. But even that did not stop them from sharing personal information.

We rely on habit in daily life because habits come to mind quickly, especially when our conscious minds are otherwise occupied or disabled. Our capacity to make conscious decisions, it turns out, is far from robust. It deteriorates under stress, it wanes when we are mentally tired, and it gets derailed by social media distractions and our own absentmindedness.

Consciousness is simply not always up to the task at hand.

*

Growing older creates disadvantages for all of us. Mental acuity, much like physical strength, declines with advancing age. Our brains show signs of this inexorable trend, including actual physical shrinkage in parts. Our capacity to flexibly navigate is affected.

A study compared younger (average twenty-two years) with older (average sixty-nine years) participants' ability to find their way in a virtual reality environment.[20] Everyone was instructed to take the shortest route possible. Participants practiced the route a certain way until they could follow it readily. Then, shortcuts opened up. Younger participants took them about 90 percent of the times they were available; older ones, only about 20 percent. Older participants, it seems, were less flexible in their thinking. They did not have the skills to quickly recognize that a shortcut would get them to their destination faster. They were reluctant to make a decision on the fly.

The decline in mental acuity is a natural part of aging. It sometimes leaves us confused. Other times, we respond only slowly. Habits are a welcome option for older people. They relieve us of having to think about how to do things and allow us to automatically act as we always have. Where are our slippers? Under the bed. Where are our keys? On the hook by the door. Where are our glasses? In the case on the table, where they've been for the past fifteen years. For an aging brain, habitual patterns make it possible to live efficiently despite reduced memory and decision-making ability.

Habit resilience illustrates an important point about the nature of habits in general: they aren't always the most effective option in a given situation, especially when that situation is complex and requires critical thought. Habits are a long-term solution, and we are banking on the prospect that their ultimate results, accumulated over time, will add up to something toward which we would not otherwise have been able to commit ourselves. Habits are what we do to get something done— because it wouldn't otherwise happen. In the present, however, and at any given moment, a habit can be a drag on your performance.

What we can learn from the demonstrations of resilience in this chapter is that we don't have to despair when we feel drained by life, when we can't seem to make excellent decisions. We can trust that parts of us will continue working on our long-term problems and long-term solutions. Likewise, the resilience aspect of habits means we have a new way to think about distraction in our lives. Distractions aren't a test of our cognitive ability. They aren't proof that we're all as flighty as houseflies. Distractions are simply an opportunity for resilient habits to surface, ones that we might choose to tamp down a bit if we had the ability to deliberate.

Habits are not malleable or creative, but they get us to our destinations in the end. Given decision-making strained by stress, tiredness, distraction, or lack of ability, the balance in our lives tips toward habits. Additional reason to establish good habits so that the habitual choice is the right choice.

13

Contexts of Addiction

Quitting smoking is easy, I've done it hundreds of times.

—attributed to Mark Twain

Bad habits, by definition, are things we wish we didn't do. But not all bad habits are equal. Nail biting is an annoying, embarrassing bad habit. Smoking is a habit that's significantly worse for our health. Substance use disorders resemble bad habits gone amok. Addiction threatens the health of the individuals involved as well as those around them. Ultimately, its ill effects harm entire societies. Most of the research on addiction has, for obvious reasons, addressed these more serious forms of maladaptive behavior.

The National Institute on Drug Abuse defines addiction as a brain disorder involving compulsive drug seeking and use.[1] Major advances in science in recent decades have shown that by using addictive drugs, we change the way our brains work, altering basic neural structures.

The substances we most often use to supply our addictions are fiendishly successful. They get right to the source of our mental functioning. Psychostimulants hijack the neural transmission of dopamine, whereas the rewarding effects of other drugs may involve distinct neuroadaptations (e.g., transmitter, receptor systems).[2] With all drugs of abuse, our brains get a jolt that creates waves of neural changes, orienting our attention to the drug, creating initial feelings of pleasure, and motivating our continued use.[3] We start to crave the addictive substance and need higher and higher doses to get the same effect. Our judgment and ability to make decisions become impaired. We find the drug difficult to resist, despite the fact that we may not like it or its long-term effect on us.[4] Drug abuse thus resembles a bad habit in that it does not depend on our conscious desires and liking for the "high" we get through use.[5] We intend to stop, but find ourselves continuing to use.

However, one significant way that addictions and habits differ is in the commitments they ask from our conscious selves. As we've seen, a true habit makes itself known to us by how our mental commitments to its ongoing execution diminish over time. Habits settle in, and we can more or less forget about them. An addiction settles in, and it takes over our lives. More and more of our waking day is spent in its thrall. More and more of our executive, agentive selves are committed to its flourishing. Some of this agentive behavior can become quite ingenious, despite its being ultimately destructive. There are forums online devoted to sharing and compiling information for the committed cigarette smoker when he or she must travel by air: which terminals have smoking sections, which airports have areas set aside before and after security. It's a whole community built around making the best decisions . . . for the health of the addiction.

The habit-like aspects of addiction may lead us into insights about its causes and possible preventive measures. This statement is not meant to replace or refute any current conceptions of addiction. It is clearly a complex and multifaceted issue, and one that requires input from many corners. Addiction commandeers multiple learning systems in the brain. It's neurological, but it's also connected to our social circumstances. It's linked to personality traits like impulsiveness. It's even partly hereditary. None of this is news. What hasn't been examined enough is its habitualness.

Many of us have firsthand exposure to substance abuse. In 2016, almost 12 percent of U.S. adults were binge drinking alcohol, 11 percent had used illicit drugs during the past month, and more than

1 percent had a pain reliever use disorder, typically involving a prescription opioid. Alcohol abuse is even more problematic in Europe, with over one-fifth of adults binge drinking at least once a week.[6] That's millions of people.

On top of that, addiction is epidemiologically different from, say, the flu. It often isn't an acute and short-term issue. The National Institute on Drug Abuse compares addiction to chronic diseases, including asthma, diabetes, and hypertension. With treatment, these diseases might improve for a while, but relapse is common.

But addiction is different still. Many substance abusers do not want to be cured. Only about 11 percent of people who have a substance-use disorder in a given year receive treatment that year.[7] Of the remainder, only about 5 percent feel that they need treatment. The most common reason given for not seeking help is not being ready to stop using. With treatment, abusers would have to give up a craving in exchange for the hard work and pain of abstinence.

For those who go through rehab, relapse is common. With gold standard treatments of psychosocial therapy plus medication, 40–60 percent of substance abusers go back to using.[8] You might wonder about twelve-step programs, which have many proponents. These do not seem to have outcomes any better than those of the standard treatments.[9] They help some people for some amount of time.

It's clear that our current approaches to treating drug abuse have not been as successful as we would want.[10] Perhaps there are other approaches to addiction treatment?

*

The Vietnam War was, among many other things, an appalling natural experiment on drug addiction. Soldiers could be drafted at age eighteen, which meant they could go to war before they could legally drink alcohol in the United States. Because of this, many of these young soldiers had not developed any kind of relationship with intoxicants before going to Vietnam. In terms of chapter 10, their deployment was a major discontinuity. One aspect of that discontinuity was that they were suddenly surrounded by a generous supply of heroin and other drugs. Heroin especially was so cheap and pure that soldiers could mix it with tobacco, smoke it, and get high. Many did.

In 1971, several years into the declared war, two congressmen traveled to Vietnam on an advisory mission and came home reporting that about 15 percent of soldiers were addicted. A *New York Times* front-page story in May 1971 titled "G.I. Heroin Addiction Epidemic in Vietnam" claimed that "tens of thousands of soldiers are going back [to the States] as walking time bombs."[11] The Army's treatment of choice was punishment, with the potential of dishonorable discharge or arrest.

The high level of drug use was shocking. It was a sensational story that found its audience. Opposition to the war had started to gather, and the public had become jaundiced toward the military and its stewards. Many people were actively protesting. Disappointment at the direction of the war and confusion about its purpose started to mix together and find root in some quarters as a general anti-military feeling. The news about addiction added to the stigma soldiers were facing upon return. The public was unnerved by the prospect of junkie soldiers overwhelming drug treatment programs, unable to hold jobs, increasing crime, and burdening welfare. Sensibilities toward addiction were raw and frightened.

In response to the threat of a drug crisis, President Nixon created the Special Action Office for Drug Abuse Prevention in 1971. Skeptics viewed this as an attempt to share the blame for the failed war—placing it on addicted soldiers. Others saw it as a progressive recognition of the power of rehabilitation; still others, as an attempt to distract public attention from Nixon's failure to obtain peace with honor.[12] Regardless, Nixon acted.

Dr. Jerome Jaffe was named the nation's first drug czar. He immediately instituted urine testing of all returning vets to determine the extent of the problem. Before being shipped home, they had to have clean tests. If they tested positive, they were sent to detox for a week or two before being evaluated again and allowed to return home.

Jaffe also decided to track what happened once the vets returned. He appointed Dr. Lee N. Robins to lead the research project. She had already made her mark as one of the first female professors of psychiatry at the prestigious medical school at Washington University in St. Louis.

Robins tracked 470 enlisted men returning to the United States in the month of September 1971.[13] More than 85 percent reported being offered heroin while in Vietnam. Forty-five percent had experimented with narcotics. While there, 20 percent claimed that they felt strung out on drugs or addicted. About 11 percent tested positive for narcotics when leaving the country. There's good reason to believe they were addicts. These soldiers continued to use right up to departure, despite the many warnings and the inevitable consequence of a week or two delay for detox and retesting.

As a Vietnam veteran told me,[14] "I used marijuana regularly. Some guys got strung out on heroin. The majority used drugs. It was so easy to do." He explained, "Once you were there, you're fighting. You had an assignment to do, and you did everything you could to survive that assignment: bring home the people in the tour. I was on a swift boat, and one of our main functions was to do insertions of troops into certain areas such as river mounts and stuff, landing them and getting them out. If I let them down . . . So I had to do everything I could to get them home. You take stuff to stay more alert. The marijuana was after, to relax."

To track rehabilitation of the soldiers who used drugs, Robins studied an additional group of 469 men who had tested positive for opioids once they returned to the States.[15] Six to eight months after coming home, they were interviewed in person and underwent a urine test. Robins was farsighted, and this follow-up is why her research made history. It's a story that is sometimes overlooked, especially given the currently popular disease model of addiction. It's one whose contours will seem familiar given the subject of this book.

Only about 5 percent of the soldiers who were addicted to heroin or opium in Vietnam—as indicated by their positive urine test on leaving—continued their narcotic addiction stateside in their first year home.[16] It wasn't because they were unable to locate a supply. About half of the formerly addicted soldiers did indeed try heroin or opium again in the States.[17] But contrary to dire predictions, the vast majority did not continue as heavy users once they returned home. Treatment didn't explain this remarkable recovery. Only about 6 percent of those who tested positive at departure even received treatment.[18]

The results confounded everyone's prior assumptions, and were highly controversial; Dr. Robins was widely criticized. Soldiers and their families were offended by the evidence of addicted troops. Rehabilitation experts were skeptical of addiction apparently dissipating so fast without professional intervention. Politicians on both sides of the aisle claimed the findings were politically motivated. The Department of Defense gladly seized on evidence that they had not consigned a generation of young men to lives wasted on heroin. A skeptical *New York Times* reporter spent two months investigating the research but then scrapped the story, apparently because there was no scoop to be had.

According to drug czar Jaffe, "Everyone thought there was somehow she [Dr. Robins] was lying [about the findings], or she did something wrong, or she was politically influenced. She spent months, if not years, trying to defend the integrity of the study."[19] Even Dr. Robins seemed surprised, noting, "Our results are different from what we expected in a number of ways." She did not like having her integrity challenged: "It is uncomfortable presenting results that differ so much from clinical experience with addicts in treatment."[20] In a retrospective article almost twenty years later, she stuck with her conclusion: "Addiction was rare and brief after return."[21] But it was clear that she still felt the need to defend the research—"I have still not found a serious flaw in the study"[22]—and to defend herself: "I am unrepentant."[23]

Her findings are no longer controversial. But in the modern world, in which addiction is regarded as a brain disease, they are often overlooked in research and treatment.

Yet the question remains, how did this happen? Why were everyone's expectations about the difficulty of rehabilitation so wrong? I think there's an untold story here about context.

Most soldiers' drug use began in the military theater of Vietnam. While there, it was easy to use heroin and other drugs. But once the soldiers were home, the context changed. There was none of the extraordinary stress inherent in a war zone. As a veteran described it to me, "You were coming from a bad place back to a good place. I didn't need it anymore because I was leaving the past behind." Back home, there were few, if any, fellow soldiers around using heroin or opium. Even the means of heroin use differed. In the States, the drug was not as pure and often required injection. With these changes, even reexposure did not lead to re-addiction. Back home, most soldiers overcame their drug use. Vets took back their lives and pursued the many other opportunities that can engage twenty-year-olds, like education, jobs, and relationships.

There's an important caveat. Drug cravings are real. For the 5 percent of soldiers who continued to use heroin, addiction was tragically all-consuming. For the rest, the results show the power of contexts to influence narcotic use, supposedly one of the most ferocious of addictions.

You might say, well, the experience of these soldiers doesn't tell us much about rehabilitation. What people go through during war has little to do with our regular lives. But that is the point. The Vietnam War context was what spurred many soldiers to use drugs. Once regularly using, they should have succumbed to the power of the drug. Instead, coming back home to different surroundings was a deterrent for 95 percent of users. When environments changed and imposed significant friction on drug use and driving forces on alternative actions, most soldiers quit.

From our habit perspective, the soldiers' return home was a significant change of context—new surroundings with restraining forces that put the brakes on heroin use. The new actions encouraged at home yielded significant rewards (a paycheck!), and the soldiers acquired new habits. For them, drug abuse was history—a bad habit that discontinuity had broken.

*

About the same time as the pioneering research on the Vietnam War, animal experiments on the same subject were uncovering new clues as well. For obvious ethical reasons, many studies of drug use are conducted on rats and not people. A lot of this work has followed the disease model of addiction, identifying how drug use changes rats' neural processes and structures. But some studies also evaluated how context affects the animals' drug use and rehabilitation.

In what became known as the "Rat Park" experiments, rodents living in different conditions were given opioids. Some were housed alone in cages. Others were kept in a colony setting consisting of a large open box with a variety of small boxes inside for hiding and nesting.[24] During the experiment, each housing unit had a drinking dispenser with a water-sugar drink and another with a morphine-sugar drink.

Where the rats were housed had a strong impact on consumption. The ones living in isolation consumed more morphine. For a social species like rats, living alone is stressful. Also, because there wasn't much to do, there were few alternatives to impede drug use. Animals living in the colony consumed less. Getting high on the narcotic interfered with typical rat behaviors of nest building, mating, and fighting. In the Park, these activities competed with drug use. Initial use is one thing (rats are curious creatures, after all), but for ongoing use—use that might resemble addiction—the context was hugely influential. Although research-savvy readers might wonder how to interpret an experimental manipulation with so many facets, including stress and competing activities, the study is at least metaphorically similar to soldiers returning from Vietnam.

The real question is, what happens once rats become addicted? Does context matter then? To answer this question, studies have tested whether rats voluntarily stop using in social contexts with driving forces that conflict with intoxication. In one, rats raised in isolation were trained to press a lever to obtain cocaine for several hours for each of fifteen days.[25] By the end of training, the isolated rats had learned the task well and were consuming a lot of the drug. For the next three weeks, no cocaine. The rats were essentially in detox. Some rats spent this time in isolated cages like the ones where they grew up. Other rats were put into a parklike colony with more rats, where they could interact and do as they pleased. Then, they were all put back in individual cages with the levers again. This time, however, when they pressed the lever, there was no cocaine. How many times would they press it now? Rats who had joined colonies pressed the levers half as often as animals kept in isolation. All rats were initially addicted in isolation, but those who moved to the colony reduced their attempts to get the drug (and presumably their craving for it), compared with rats that continued to be isolated.

This power of contexts suggests that substance abuse is partly an adaptation to environmental circumstances. That is, addiction doesn't arise just from past drug use that co-opts our brains. Instead, the researcher behind Rat Park, Bruce Alexander, argued that it's an attempt to cope with current circumstances—ones that have few restraining forces on use and that offer few other rewarding activities.[26] The ramifications of this idea are enormous. It shifts the location of the dysfunction from people to the environments in which they live. It treats addiction as an adaptation, just one that happens to come with serious health problems, disruption to loved ones, and social stigma.

We can see why a drug addict who's living primarily on the street isn't as successful at kicking the habit as were returning Vietnam soldiers (or rats in their park). The return home created a discontinuity of time and place between the environment where the habit was first picked up and the environment where the (former) user continued to live the rest of his life.

Similarly, drug abusers who are admitted to a standard inpatient recovery facility are away from home and no longer exposed to the environment that permitted and encouraged their drug use. While a patient, they dry out, receive psychological and medical treatment, and are engaged in different activities. This environment makes drug use much more difficult than it was out in the world. Inside a treatment center, it's not so hard to quit. Once treatment ends, however, they return to where they used drugs habitually. It's no surprise that (1) they are able to kick the habit in a radically new environment, and (2) 40–60 percent of users relapse after treatment, after returning to the old environment.[27]

Take the experiences of thirty-two Australians who had been

treated for alcohol and opioid addiction.[28] They were interviewed once a year for three years following treatment. Most believed they had average to strong willpower, regardless of whether they were successfully abstaining.

Three years later, only five of the thirty-two had stayed off drugs completely. This handful in stable recovery stood out in yet another way. They had made radical changes in their living arrangements. Some had given up public housing support, moved to a new city, and gotten jobs where no one knew they had been addicts. One moved in with his girlfriend, who did not use. The only one of the five who did not move away had switched his group of friends and gotten a full-time job that made drug use difficult.

The other twenty-seven respondents, the ones who didn't abstain for the full three years, had not made such changes to their environments. Many blamed relapses on living situations that made it easy to use drugs and alcohol. They continued to see friends who used. They were still exposed to people selling addictive substances. As the researchers concluded, "The most important difference between those who have achieved meaningful recovery and those who have not lies not in their skills or knowledge, but in whether they were able to overcome the financial and social obstacles to moving to a nonpathogenic environment."[29]

In this model of drug use disorder, friction accounts for so much. If we make the object of attraction less available, and if we move the person out of a context rife with cues of use, that person will use less. Admittedly, drug use is complex, and to rely on external forces alone is surely overly simplistic. But this understanding is not just simple; it's also humane. It does not locate a brokenness in the mind of the user, at some deep core of who they are. Instead, the danger they're inviting into their lives is on a continuum with all the minor dangers we all invite in as well, dangers that the world makes easy and abundant.

*

The more conventional model for drug use is that addiction is a chronic brain disease characterized by compulsive drug seeking and use. I don't want to overstate the differences between this disease model and the context model. In reality, these are two sides of the same coin. Our brains respond to rewards in our life contexts, especially to drug rewards. In turn, how we experience and process rewards depends on our neural systems. In this way, our contexts reach deep into our minds, it's true. And that's where the alchemy of addiction takes place.

Where the two views really diverge is when it comes to how we treat addictive dependence on drugs.

The disease approach addresses the cognitive, affective, and neural cravings initiated by drug abuse. It medicalizes drug abuse. As a choice of mitigation strategies, the disease model therefore attacks the drug itself. Prohibition is the most common form this takes. Law enforcement seeks to interrupt and stop the flow of the substance, under the assumption that the drug itself is the primary cause of addiction. Treatment also can involve detox meds, such as lofexidine for withdrawal symptoms in opioid dependence.[30]

The success rate of this approach is not impressive. Again: the National Institute on Drug Abuse's own estimate is 40–60 percent relapse.

Why can't we shoot for 5 percent? That's the relapse percentage of the returning Vietnam soldiers. Of course, that wasn't a controlled experiment, but there's no reason why we shouldn't use that as a benchmark of success (and possibility).

What would rehabilitation look like if we took the implications of Vietnam and Rat Park seriously? Treatment in this alternative model would build on the recognition that drug abuse is especially prevalent in impoverished environments with few competing sources of alternative reinforcement and many cues driving use. It would recognize that those with low income and education levels in the United States are most at risk of abusing cocaine and opioids. It would focus on changing, or destroying, the environments of abuse.

This idea already has proponents. An article in a top substance abuse journal argued that current treatments are largely unsuccessful because they "focus too much on reducing substance use and not enough on linking clients to reinforcers that will make abstinence more appealing."[31] To get people to consider treatment requires "sufficient incentives in the environment to make the effort needed to sustain long-term abstinence worth it."[32]

People have been making this argument for decades. The problem is that this approach has always taken a back seat. A classic treatment from the 1970s proposed to change environments in just these ways. The *community-reinforcement approach* was designed specifically to make abstinence more rewarding than drug use.[33] The treatment does not remove people from environments so much as change what cues and reinforcements are available where they live. As originally conceived, it was highly intensive, with many parts, including therapy, employment counseling, couples counseling, and help in forming new social networks.

Most tests of this approach have adopted only some of these components, and few have evaluated long-term effects. Even more unfortunate, we lack critical information on how to make abstinence rewarding. We know little about the kinds of non-drug rewards that keep people sober. Everyone responds to money in the short run, but abstinence has proven difficult to maintain when payment stops.[34]

Some evidence suggests that social rewards are important, just as they are in our personal habit formation. One study changed the social networks of alcohol-dependent people by encouraging them to find friends who disapproved of drinking along with activities that did not involve alcohol.[35] Attending AA meetings was proposed as a way to meet new people who were abstinent and enjoyed activities other than drinking. The researchers downplayed other aspects of AA. With this change in social reinforcement, participants were drinking less even two years later. Forty percent of participants reported complete abstinence after two years, compared with around 30 percent of a standard treatment group.

Employment-based rewards also are relevant. "Therapeutic" workplaces give unemployed drug-dependent individuals job training and employment. This innovative treatment was commended in 2014 by the White House Office of National Drug Control Policy. Employment opportunities mostly involve basic computer skills, with drug-free urine samples required in order to continue work and obtain maximum pay. A review of eight such interventions revealed significant reductions in opioid, alcohol, and cocaine dependence during treatment.[36] In one study with cocaine-dependent welfare recipients, about 80 percent of tests were drug free during the eighteen-month workplace employment treatment.[37] However, once the program ended and rewards ceased, participants were back in their old life contexts and mostly resumed old patterns of use. Perhaps the programs simply did not last long enough to inculcate skills in the participants that would then, upon return to their normal lives, allow them to bring about their own discontinuities by finding new jobs, new communities, new possibilities.

Treatment models that change the environments of drug abuse by offering new reinforcements for sobriety might seem a costly alternative to the current disease model. However, policy implementation could take many forms, including government wage supplementation for abstinent workers, alliances with cooperative employers, and the creation of new therapeutic businesses designed to provide sustainable employment.[38] The feasibility of large-scale urine drug testing has already been demonstrated with Department of Transportation employees, who regularly undergo such screening.

*

A postscript about the role of habits in addiction is useful. Most often, habits are discussed as part of the dysregulation of neural circuits, especially circuits involved in liking and wanting the addictive substance.[39] But habits play a more felicitous role, too. Under the right circumstances, habits support recovery.

John Monterosso and I interviewed eighteen members of twelve-step programs, each of whom had achieved more than two years of sobriety.[40] They identified behaviors that they believed were important to keeping sober. The most critical things, they said, were: going to meetings, practicing gratitude, being rigorously honest, keeping busy, practicing prayer, acting in their role as sponsor, and helping others. Participants said these actions were the keys to their recovery. In another part of the interview, they rated the habit strength of each of those behaviors, indicating whether they did them automatically and without much thought, or only after considering other options. What's most interesting is that importance and automaticity were closely related. The more *important* the behavior was to their recovery, the more *habitual* it was in their lives. These individuals seemed to understand their propensity to respond out of habit. In turn, they had practiced recovery behaviors to the point at which they became automatic. They had replaced bad habits with good ones. They had adapted some of the same neural mechanisms that promote habitual drug use to instead support their well-being and sobriety.

For drug treatment, new ideas are clearly needed. We've seen limited success from current treatments but much pain, wasted lives, and enormous costs from drug abuse. Maybe it's time to put more emphasis on the behaviors and contexts of addiction and recovery.

14

Happy with Habit

[People] become builders by building and lyreplayers by playing the lyre; so too we become just by doing just acts, temperate by doing temperate acts, brave by doing brave acts.   —Aristotle (translated by W. D. Ross)

If you want to get a young child to eat vegetables, what do you do? We expect that children have strong natural preferences that attract them to food that's bland, sweet, or fatty, like milk, cookies, pizza, hamburgers, hot dogs. So you might think to add a bit of sweetness to the vegetables before serving them, perhaps making glazed carrots. Or you might add some fat, perhaps whipping butter into mashed potatoes. We'd predict that kids would find them more palatable these ways.

But there's an even simpler way to get them to eat more healthfully: just keep trying.

A group of researchers in the U.K. tracked the meal habits of toddlers in a preschool. Every two or three days, the preschoolers got foods they would typically refuse: snacks of pureed carrots or artichokes.[1] As you can imagine, initially, artichokes were not wildly popular. Most kids had never seen or tasted them before. Neither had many of the nursery staff, who had to be instructed not to make disparaging comments or unpleasant facial expressions. Kids can see through a poorly executed ruse.

The initial attempts weren't all that successful. The kids started off eating a little more than an ounce. They took barely a taste. During the next two months, each child was given artichokes about fifteen times. Each time they got the stuff, they ate a little more, and then a little more. The biggest gains were during the first five tries, and then the increases leveled off. By the end of the study, more than five ounces of artichoke per person were consumed on average. That's quite a bit of artichoke, especially when you're a human who weighs approximately forty-five pounds.

Maybe the kids would eat even more of the vegetables when they were made more palatable? To test, researchers gave some children slightly sweetened artichoke puree. Others got fat added, to give the vegetables that smooth texture kids like. But these modifications made no difference in how much the children ate. Kids ate more of the artichokes with more exposure to them, not with more sugar. Not every kid, of course. Sixteen of the seventy-two were holdouts, never learning to eat vegetables. They never got past the look and smell. For the rest, however, repeatedly having artichokes got them to eat more. We can't know for sure what they would have stated as their preferences—perhaps they would report being just as artichoke-skeptical at the end of the experiment as they were at the beginning. But the point, after all, wasn't about claiming to like vegetables; it was about eating them.

When you think globally, the artichoke findings are not surprising. Kids eat all kinds of things when exposed to them often. In Japan, breakfast is rice and a fermented soybean paste. In China, kids eat *jook*, a rice porridge topped with strings of dried meat, egg, or pickled tofu. In Latin America, even very young children drink coffee with milk. In Mexico, they dip their tortillas in a bowl of hot salsa or enchilada sauce. Children come to eat foods that are sour, fermented, and spicy, if that's what they get.

Adults are just as impressionable, although we don't usually recognize it. Yes, we repeatedly do the things we love doing. But we also grow to love the things that we repeatedly do. It's like an invisible feedback loop inside our heads. As you can imagine, this loop has something to do with our habits. And it has a lot to do with our happiness.

*

In 1910, a psychologist named Edward Titchener observed that well-known objects, merely because we have seen them before, cause us to feel a "glow of warmth, a sense of ownership, a feeling of intimacy, a sense of being at home, a feeling of ease, a comfortable feeling."[2] We like things simply because we have grown used to them, he maintained. In a 1968 paper, the social psychologist Robert Zajonc called this phenomenon *mere exposure*.[3]

There are many reasons why exposure leads to liking.[4] One of these is *familiarity*. Used loosely and commonly, that word just means that we recognize something we've encountered before. But sometimes the deeper meaning peeks through—we might acknowledge that we prefer something for no other reason than that it feels familiar. There's something substantive at work there. Familiarity explains why photographs of our faces often strike us as odd. It can seem as though you're looking at someone you don't really recognize. The reason is that your face is not completely symmetrical across the midline. The left and right sides to your face aren't exactly identical. The face that the rest of the world sees—the face that's captured in a photo—is the reverse of the face we see every day in the mirror. For most of us, the asymmetry is enough to be detectable, and it makes photos of ourselves seem unfamiliar.

A clever experiment showed college students two pictures of themselves: one that was their true photographic image, and another that was flipped, as they would see it in a mirror.[5] Which did they prefer? Students liked the mirror image—the one they've been staring at all their lives. Then the researchers showed both photos to the students' friends. The friends liked the face to which they had grown accustomed—the photographic image. In both cases, preferences had more to do with repeated exposure than with aesthetic quality.

The exposure effect also signals welcomed *predictability*. If you travel a lot for work, you probably have a favorite go-to chain restaurant.

Travel is easier if you automate basic decisions like where to eat. Of course, even the best chain restaurant won't be as good as a great local place. But then again, it won't be as bad as the worst local place. Even though most chains don't have the best food, I bet that you have started to like your usual dinner spot. You might even stop by occasionally when you are at home. You walk in, and it feels familiar. You see the menu, and you know what to order. You can probably come up with a number of perfectly valid reasons why you like the place. Maybe you enjoy a particular salad dressing or crackerbread. But is that really why? Familiarity and predictability are certainly at play, too.

There's something else that happens as we repeatedly do the same thing: our experiences become *fluent* and easy on the mind. With repetition, we have little difficulty understanding and evaluating what is happening. Some researchers argue that fluent processing is itself enjoyable and makes us come to like repeated experiences.[6] Aesthetic preferences for art and music are enabled by *perceptual fluency* from recurring features (e.g., rhyme, melody, symmetry). Modern and contemporary art often lack such recurring features, but they have a kind of *conceptual fluency* in which ideas spur recognition of meanings and emotions.[7]

Our preference for repetition is sometimes surprising. We all think we'd love to drive around in a distinctively styled car—the kind that makes other people look twice. But when consumers rated how much they liked 3-D renderings of seventy-seven models like the ones above, they preferred the cars with conventional, typical-looking features.[8] The cars with more typical features had higher sales, too. It makes sense that futuristic car manufacturers like Tesla have stuck with standard automobile features, despite the innovations inside. Our preference is for what we are used to seeing.

We grouse about ads' ubiquity, but advertising and branding deliver to our doorstep (and screens) a whole lot of what we want to see and what makes us feel good. Consumers' affection for brands increases with greater exposure to their ads, reaching a maximum in lab research at about ten exposures, which is when habituation seems to set in.[9] Logos and trademarks mean little on their own. Through experience, we come to recognize and appreciate them. Even children are susceptible. Four-year-olds were given a chicken nugget in a wrapping with a McDonald's logo and an identical piece wrapped in white paper.[10] They tried each and indicated which tasted best. They did the same with hamburgers, fries, cups of milk, and baby carrots, comparing one item in a McDonald's wrapper and an identical one in a plain wrapper. Kids thought four out of the five items tasted better in the McDonald's wrapping. Even the carrots tasted better with the logo. Kids who ate at McDonald's more often had stronger preferences for the branded foods. The logo gained positive meaning with kids' repeated experiences at McDonald's.

Efficiency is another reason we like repeated experience. I take advantage of this in the classes I teach. Students typically just stick with the seat they choose on the first day of class. I take a picture of them in their seats and memorize their names in that order. It makes life easier for them and for me.

In explaining why they return to the same seats, students told a researcher things like: "I think the first choice was spontaneous and then I got used to that choice"; "Most times the first choice is random, and then I go back to that same place"; and "Hard to explain the first choice, afterwards it is inertia."[11] Their choices might have been accidental initially, but quickly became the default option. When asked to rate possible reasons, students reported feeling more comfortable in their regular seat, more confident and in control, and able to concentrate better.[12]

Past exposure signals *safety*, too. Residents of Edinburgh were polled about how safe they felt in areas of the city they frequented, as opposed to sections where they rarely went. When the researchers compared residents' ratings against the prevalence of crime in each area, people had an exaggerated idea of how safe they were in the places they went often and a more realistic view of crime elsewhere.[13] Put another way, the more familiar they were with a place, the less accurately they judged its safety—just the opposite of what we might expect. Our comfort with what we know can distort our sense of reality.

This also explains our driving attitudes. My sister lives in Montana and is completely comfortable behind the wheel there. She feels that she is taking her life into her hands, however, when visiting Los Angeles. In reality, her home state has topped the nationwide per capita driving death charts in recent years,[14] while the motorist fatality rates in California are relatively low. Across the country, roadside death rates are lower in urban areas than in rural ones. It's easy to misjudge safety by basing it on feelings of familiarity.

Mere exposure happens without our realizing it. When we repeat actions, our preferences change. The effects are subtle and not always apparent to our conscious minds. We think that we make decisions to act, not that our actions influence our decisions. When acting habitually, we often believe that we are acting on our desires—what we wanted to do all along. People with stronger habits of riding the bus, purchasing fast food, and watching TV news in one study were highly certain about their intentions to do these things, despite the fact that their intentions were *epiphenomenal*—noncausal, incidental. It didn't matter what they intended; they just continued to act out of habit.[15] It makes sense that we'd take personal responsibility for our habits. After all, our repeated actions feel familiar, predictable, fluent, and safe.

Throughout this book, we've talked about how to create new habits that will improve our lives. We've seen how to choose a behavior that is rewarding and then strategically change our surroundings to make it easy to repeat. In so doing, we are changing our habit self so that it becomes aligned with our conscious self in achieving our goals.

But now we know that there are many routes to that harmony. Simply by repeating actions, our desires change. We start to prefer the things we experience over and over. They become what we want to do. Habits, it turns out, are a two-way street. They achieve our goals and they become our goals, too. Do you know that immediate sense of comfort and rightness that we feel when we return home after a trip? This is just an acute experience of the aura that habits accumulate.

*

A good friend of mine is a happily practicing Catholic. She finds the rituals of attending church and participating in mass uplifting and comforting. The regularity of the sacred times, spaces, and objects provides structure. The gestures, music, Communion, and incense are reminders of the symbolic and emotional meanings of the liturgy. It is "out of the context of concrete acts of religious observance that religious conviction emerges."[16] This famous quote from the anthropologist Clifford Geertz captures the spiritual meaning that emerges from rituals.

All rituals are grounded in repetition and rigidly fixed action sequences.[17] But they differ from habits in one important way. Rituals lack a direct, immediate reward. Instead, we have to invent a meaning and impose it on them. We lift our glasses to toast, blow out candles on a birthday cake, and wear caps and gowns at graduation. The act of standing silently for a song, singing while candles burn, or wearing a ceremonial costume acts as feedback, reinforcing our belief that something meaningful is taking place—an act of respect for our country, a celebration of another year, or an educational accomplishment.

Rituals are a universal human impulse. Native Americans, especially in the Southwest, had rain ceremonies. Japanese have the art of the tea ceremony. Aztecs performed human sacrifices on top of their pyramids. To an objective eye, these rituals are not especially rational (and certainly not all are desirable). But researchers are discovering a logic behind them, especially in times of uncertainty and anxiety. Repetition is its own reward—which is something any six-year-old might consider obvious, after her fourteenth viewing of *Moana*.

Consider the high-stakes, high-pressure world of elite athletes. At the top of each sport, all players are highly skilled. A great deal of money, fame, and talent are on the line each time they compete. Winning requires a lot of confidence and some luck. No surprise, then, that sport is rife with superstitious rituals. Players use them to gain a sense of control in this highly unpredictable environment.

Thirty years ago, there was little that was flattering or functional about the fashion for long, baggy basketball shorts. They originated when Michael Jordan had to wear an extra-long Chicago Bulls uniform in order to hide his "lucky" University of North Carolina blues underneath. Now those shorts are ubiquitous. A fashion statement from a superstitious ritual! In that case, repetition found its meaning well after the fact. Repetition is powerful in that way.

Many pro football and hockey players sport lucky beards. This trend apparently started with Björn Borg, the Swedish tennis star, who repeatedly won at Wimbledon while declining to shave (and wearing the same Fila shirt). He took five straight titles.

Given the pressures that athletes are under, it is no surprise they believe in these quirky acts. Eighty percent of pros report having a superstitious behavior they perform before playing, which in one study ranged from always eating four pancakes to seeing the number 13 at least once.[18]

Mere belief is itself pretty powerful. Placebo pills can achieve the same results as actual medications if we're convinced we're getting the real thing. Our simple belief in lucky socks could actually enhance performance on the field. But there's more to this than placebo. Geertz was right about the importance of concrete acts. Ritual practice and repetition of actions have soothing qualities.

In one study, college students practiced at home for four days a set of elaborate "action sequences," such as making fists and turning them over, taking three deep breaths, and closing their eyes.[19] They learned by following a video model and written instructions. The full sequence took several minutes to complete.

The central question was, would this arbitrary ritual help students cope with failure? On day seven, participants showed up in the lab. Some performed the ritual. Others did not. Then they all completed a difficult computer detection task. To make sure they were trying hard, they were offered a $10 bonus for accuracy. Still, participants made errors about 20 percent of the time.

The study had a unique way to measure reactions to failure—electrical neural signals recorded with an electroencephalography (EEG) machine. This is a cap with lots of protruding wires that attach to the scalp noninvasively. It assesses electrical activity within the brain's neurons, in this case when students made errors at the task. On making an error, our brain exhibits a waveform called *error-related negativity* (ERN).

Students who performed the ritual before doing the task had smaller ERNs, which meant they were responding less extremely to their errors. Rituals seemed to buffer students against distress at failing. Although they did not react as extremely to mistakes, their performance at the task did not suffer (nor did it improve).

Students' reports after the task were revealing. One wrote, "The repetition of activities somehow improved the completion of the (computer) tasks. I think that maybe completing the set of actions helped me feel a little more focused and calmed." Another wrote: "Completing the actions before beginning the task helped calm myself and make me feel in control for some reason." Just performing a ritual, it seems, calms fears and anxiety. The repetitive actions may satisfy our need for order and predictability. Rituals might also distract us, blocking negative thoughts and stopping us from ruminating. Some rituals with symbolic value could directly buffer us against threat, as religious rituals provide a sense of meaning beyond ourselves. Even secular rituals could remind us of meanings beyond the threatening domain.

Many of us understand the benefits of ritual repetition. Brazilians, for example, use simple rituals, called *simpatias*, to address everyday problems.[20] In a study, U.S. college students, along with Brazilians who themselves used the rituals, rated how effectively twelve *simpatias* would deal with problems such as quitting smoking, lack of friends, infidelity, and depression. As examples:

"Wear a white T-shirt for five days in a row. After that, wash the T-shirt using salted water. Put the T-shirt to dry in the shade. After it has dried, fold the T-shirt and take it to a church."

"In a metal container, put the leaves of a white rose. After that, set fire to the leaves, get the remaining ash from the leaves and put it in a small plastic bag. Take the small plastic bag and leave it at a cross-road. Repeat the procedure for seven days in a row."

Obviously, these actions have no real power to alter lives in a magical way. That's not what's interesting. Consider the fact that these rituals have probably existed for a very long time, or have at least been repeated orally and spread to a great many people. Have they ever worked? No. Not once in actuality (coincidences notwithstanding). So effectiveness is not making these spread—it's our beliefs. The fact that a *simpatia* is repeated gives it a kind of special power per se, regardless of its results. Both Brazilians and Americans expected that repeated actions and greater numbers of actions would be more effective. As you might guess, college students were skeptics in general about the efficacy of any of these compared with Brazilian practitioners.

Of course, the reality is that an ineffective action is ineffective, whether done once or ten times. But even in our intuitive beliefs, we favor actions that we repeat in the same way each time. When pushed, we might even accord them a bit of magic.

Before you start to think that such rituals are a part of some exotic Brazilian culture, I'll note that an online survey asked Americans about any rituals they had developed after the death of someone close to them or after the end of a relationship.[21] Responses included:

"In these fifteen years, I have been going to hairdressers to cut my hair every first Saturday of the month, as we used to do together."

"I returned alone to the location of the breakup each month on the anniversary of the breakup to help cope with my loss and think things over."

"I looked for all of the pictures we took together during the time we dated. I then destroyed them in two small pieces and then burnt them in the park where we first kissed."

"I washed his car every week as he used to do."

It seems that rituals help people cope in times of stress and loss. They become the reassuring, familiar actions that bring peace and calm. It's not only loss that spurs ritual formation. Almost half of all people in an online survey had developed rituals to use in times when they faced a difficult task and felt anxious about it.[22] The rituals were typically repeated activities and were rarely constructed on the spot. Repetition is characteristic of the rituals in our lives.

To try to pin down what makes rituals work, a lab study with college students tested how people cope with losing money.[23] Students met in groups of about twelve. One person in the group was to win a lottery of $200. To heighten the other students' desire to win and angst at losing, each had written, before the selection, about how they would use the money if they won. A single winner was randomly chosen and excused from the study (yes, that person got $200!). The remaining participants were told that people often engage in rituals to help them cope with a loss. Some of the students were instructed to complete a ritual with many steps—drawing a picture about how they felt, putting salt on it, tearing it up, and counting to ten five times. Participants who engaged in the ritual activities reported less grief over losing and higher feelings of being in control than the participants who were only told about rituals. Performing the ritual was what seemed to matter and to reduce the anguish of not winning the money.

A similar draw-salt-tear-count ritual also worked to reduce performance anxiety. College students in another set of studies were informed that their task was to sing "Don't Stop Believin'" by Journey before an audience.[24] Some participants were given time to try to calm themselves down before singing, whereas others used the time to perform the ritual. Participants armed with the ritual reported less anxiety, their heart rate stayed lower, and they sang the song more accurately (volume, pitch, and note duration scored by a karaoke program). Just trying to stay calm did not have such beneficial effects. It might seem surprising that a single ritual performance had calming effects, but the "ritual" label seemed to be important. When students performed the same activities described as just a set of behaviors, without the label of "ritual" and its implication of repetition, their anxiety did not subside.

Sports fans who love their teams also have to cope with loss and anxiety. No surprise, then, that many have superstitious rituals. In one study, about 40 percent of college students reported that they engaged in a ritual to help their team.[25]

The most common superstition was wearing the right apparel. Examples were: "Wear lucky jersey I bought when they beat the NY Mets by six runs." And "Wear the jersey. If Pats are losing at halftime, take it off." Nonalcoholic drinks and food were also important: "When we go, we each have a specific food that we must eat during the game." "If I eat grapefruit for breakfast, they win." Fans reported additional odd rituals, including a woman who reported that during a national soccer tournament, "Legs are not to be shaved." A Houston Aeros hockey fan: "Before every game I put my socks in the freezer for two hours and then wear them to the game. . . . This is what they do to the game pucks. I feel it gives us a slight advantage." The reasons for these? Fans explained, "I'm doing my part," "I can help the team out," "On notable occasions it has affected the outcome," and "Pull off the win through my good karma." Superstitious rituals seem to give fans, just like the sports stars they are rooting for, a sense of control over the outcome.

It's easy to trivialize rituals and believe they are silly superstitions that have no effect. But in stressful times of uncertainty and loss, enacting practiced sequences of behavior helps us cope with our feelings and gives us a sense of control—even when our discomfort as a fan wearing frozen socks could never actually make a difference.

*

Our inferences about our own habitual behaviors make them seem normal and reasonable—to us. But as Mark Twain is said to have quipped, "Nothing so needs reforming as other people's habits."

People who squeeze the toothpaste tube from the bottom find themselves making well-reasoned arguments about why middle-tube-squeezers are doing it wrong. Or maybe you hang the toilet paper rolls in one particular way. You come up with convincing reasons no matter which way you choose. Random patterns become habits simply because they're the way we have always done it. Might sound silly, but it feels true.

The positive feelings created by acting habitually have even broader effects, enhancing well-being and meaning in life. To most people, life meaning is a lofty notion tied to spirituality, love, and great accomplishment. But habits provide a quieter bedrock of such meaning. The right habits are an often unrecognized launchpad to experience the positive mental state of *flow*, or the focused enjoyment you feel when immersed in a skilled activity.[26] My husband, for example, persistently practiced his fly fishing casts to Hula-Hoops in our backyard. Now he gets out on the water and loses track of time and the beating sun in a sort of fishing high. For you, the right habits could fuel a passion for music, for writing, or for creating in the kitchen.

Just acting habitually has broader effects of reducing uncertainty and promoting feelings of coherence and comprehension of our experience. In a survey of daily routines, people who reported that they do "pretty much the same things every day" found life more meaningful.[27] This was also true moment by moment. When contacted during their day, people reported more meaning in life when performing actions that were part of a routine. As one of the study authors, Samantha Heintzelman, noted, "The applications sort of jump out."[28] Life meaning can come from maintaining a tidy office, keeping a daily schedule, having weekly dinners with friends, or walking the same path to work or school every day. This is the coherence of an ordered life. And it's a coherence attainable by all of us.

This simple effect—from repetition comes liking—contributes to habit persistence. When habits are beneficial or even neutral (really—toothpaste?), liking them works in our favor. Our good feelings reconcile us to the dailiness of our lives and persist long after we have habituated to any rewards. These inferences are beneficial when we gain appreciation of our saving, exercising, and productive-work habits and so come to value them more and more as we repeat them.

You see this basic phenomenon play out in many ways. Older people interviewed about their favored products revealed the expected lineup of Pond's face cream, Tide detergent, and Heinz ketchup.[29] Although these shoppers easily named often-used products, few had a ready explanation for their preferences. The interviewer concluded that "whatever prompted them to start buying a product held less importance than their current comfort level and sense of familiarity with it." Through our actions, "the best thing and the thing we are most comfortable with, maybe, could be virtually indistinguishable."

But there's also a cautionary note to this tale. We may wind up embracing repeated actions that are not good for us. Through exposure, we can become reconciled to behaviors that are not ideal. We keep procrastinating, eating too much, exercising too little, because that's what we've always done. We persist with little reason except for the pull of previous repetition. We end up liking even our maladaptive habits. What we learn from research on exposure is that this liking will subside only if we form new habits that themselves become, through repetition, the familiar and comfortable.

You Are Not Alone

Sometimes it feels like this. There I am standing by the shore of a swiftly flowing river, and I hear the cry of a drowning man. So I jump into the river, put my arms around him, pull him to shore and apply artificial respiration. Just when he begins to breathe, there is another cry for help. So I jump into the river, reach him, pull him to shore, apply artificial respiration, and then just as he begins to breathe, another cry for help. So back in the river again, reaching, pulling, applying, breathing and then another yell. Again and again, without end, goes the sequence. You know, I am so busy jumping in, pulling them to shore, applying artificial respiration, that I have no time to see who the hell is upstream pushing them all in. —John McKinlay, epidemiologist

To illuminate the force behind our habits, it has been necessary to show how drastically most of us overestimate the strength (and necessity) of conscious thought. It's not that it's ineffective; it's just not as tenacious as we think. We can blame our egotism for the distorted self-perception: to our conscious minds, each of us is uniquely in control of our lives and behaviors. So when we fail to achieve our goals, we feel doubly cursed: not only did we fall short of the finish line, but we also never had the stuff to succeed in the first place. As we have seen, we don't need to feel this double failure. Our habitual selves can take on much of the steadfast drudgery needed to achieve the goals set by our conscious selves. It's a more efficient and happier way to live.

But there are still some things that our habitual selves can't solve—because there are some things that are simply too large for any one person to tackle alone.

Many of your own challenges are just not, well, all that personal. Look around you. Forty percent of Americans are obese, half of marriages end in divorce, and people retire with an average of $17,000 in savings. These horrendous statistics hide a liberating insight: your health and welfare are not just your own personal responsibility. They are socially shared challenges that reflect the trials we all experience in the broader environment in which we live. This has implications for how we think of habits, and how we organize better environments for all of us, together. By looking upstream, you can identify the forces that push us all into streams of trouble, as John McKinlay claimed. You have nothing to lose except your lingering sense of failure.

*

The Nobel Prize–winning economist Richard Thaler and his coauthor Cass Sunstein coined the term *nudge* in a 2009 book on social policy. They were approaching it from a top-down perspective, from the field known as *behavioral economics*, but their conclusions are useful here. Much as we have discussed cues and contexts at a personal level, they showed how changing the *choice options* in our collective environments is part of smart social policy. Taxation is the prototypical policy for changing behavior. But as they pointed out, policy solutions don't need to be coercive like taxes, which most of us pay only grudgingly. Instead, Thaler and Sunstein proposed policy changes that involve a nudge, which "alters people's behavior in a predictable way without forbidding any options or significantly changing their economic incentives."[1] Their work is immediately intelligible to you and me as controlling the driving and restraining forces in our environments.

Take organ donation. In many countries, this is an *opt-out* decision. People are assumed to have given their consent unless they decide not to donate their organs upon their death. As you can guess, opt-out countries such as Spain, Austria, and Singapore have highly success-ful organ-donation programs.[2] In the United States, we still have to *opt in*. You are assumed to have rejected organ donation unless you explicitly say otherwise, usually by checking a box, say on your driver's license application. The result is an organ shortage in the United States. More than one hundred thousand patients are on waiting lists to get transplants. Many will not live long enough to get one. The opt-out approach is in tune with what we know about the greater demands of conscious decision-making versus habit's efficiency. Deciding to do something—like lose weight or donate your liver—is much easier when your choice options are structured in a way that facilitates your behavior.

Other nudge policies involve simplifying information, giving warn-ings, and highlighting social norms. These, as we have seen, may not be strong enough to alter already-formed habits. But there's power in rearranging our environments to make desired actions easier. A well-known nudge is Thaler and Shlomo Benartzi's Save More Tomorrow program.[3] In the past, you had to decide to join your company's retirement-savings program and decrease your paycheck today in favor of investment for your future. Now many businesses automatically enroll you, as a new employee, in these retirement plans, which link your savings rate to future pay increases, so your take-home doesn't go down (reducing *that* friction). The plans are opt-out. To decline, an employee has to actively decide not to take part, filling out a form say-ing, in essence, "No, I'd rather spend the money today than save it for my old age." Showing its success, when Thaler received the Nobel Prize in 2017, the program had already increased retirement savings by about $29.6 billion.

It's a brilliant idea to base social policy on science. Our conscious selves underestimate the impact of external forces in our everyday con-texts, and science-based policies can correct this. Great Britain cre-ated a Behavioural Insights Team to use science in this way. They design government policies to change everyday environments so that it's easier for people to make good choices (www.behaviouralinsights .co.uk).

The United States, as always, is a bit of an outlier where it comes to policy. We now have a federal Social and Behavioral Sciences Team, but its influence is far less than that of its counterpart in the U.K. Independence, rugged or otherwise, is still a hugely seductive idea in the United States. It's not that we don't like to help one another; it's that we still tend to assume that self-control and willpower are the only authentic ways to achieve results. In reality, we're just making life more difficult for ourselves, and setting up the vast majority of us with normal levels of willpower to experience failure.

Still, there are wonderful semi-experiments happening all across the country. The great diversity that is America comes into play here. The country is a loosely knit federation, with each state and town having to some extent its own laws, values, history, and economy. This diversity allows us to compare areas of the country where people are better able to achieve common life goals with areas where they fare worse. Quite simply, people are already acting in ways that make them healthier, wealthier, and happier in some areas of the country than in others. Of course, we can't know for sure what, exactly, creates more beneficial habits and lifestyles in one place than in another (the reason I called these "semi-experiments"). But as you will see, we can often make good guesses about the kinds of social policies that could change the forces in our living environments to enable more of us to achieve our goals.

*

If you asked me, "What can I do to exercise more?" by now you'd expect a spiel about how to set driving forces and remove restraining forces to exercise repeatedly, along with the importance of rewards. You'd expect me to explain how you, personally, can form an exercise habit given your own circumstances. That's a good answer at one level. But there's another answer in the shared driving and restraining forces that are the default options for all of us in our environments.

In some places, people exercise more often than in others. More than 25 percent of the residents of Colorado, Alaska, and Washington, D.C., in 2014 met government recommendations of 150 minutes a week of cardio and twice-weekly resistance training.[4] No surprise, then, that residents of Colorado and Alaska had the lowest prevalence in America of type 2 diabetes, and Colorado of hypertension as well.[5] Washington, D.C., wasn't far behind on both measures.

Those exercise numbers were halved in Tennessee and West Virginia, with less than 13 percent regular exercisers. In some states, many people didn't even try—a third of the residents of Alabama, Louisiana, and Mississippi didn't exercise at all. In conjunction, residents of these states had some of the highest rates of disease. These states all were in the top ten for type 2 diabetes and high blood pressure.

What is the magic ingredient that separates healthy and unhealthy states? One answer is the people who choose to live there. People who like to exercise move to the outdoorsy states such as Colorado and Alaska that project a rugged image of skiers, rock climbers, and kayakers. Washington, D.C.'s image is active urban, with pedestrians, cyclists, and joggers. In contrast, Louisiana and West Virginia do not evoke images of active lifestyles. More sedentary types are likely to feel comfortable there. One should never underestimate the power of human self-sorting.

Yet, another answer is that local programs, culture, and politics influence the behavior of residents of each state. In Colorado and Alaska, for example, the outdoor recreation industry is a dominant force. And then there's your neighbor's behavior. If you live in one of these states, your neighbors are likely to invite you out for a jog, your kids will bike to their soccer games, and town residents walk to the store. At a certain point, peer pressure kicks in. But even before that, you'll simply be selecting activities from a different menu. If you live in a place with more sedentary neighbors, you are more likely to get together for dinner or a card game than pickup basketball.

There's more here than a simple academic analysis. What you really want to know is what would happen to *you* if you moved to a more exercise-friendly state. Could something about your neighborhood really improve your own fitness and health? Would this happen just . . . well . . . magically? Would you weigh less, too?

Of course, I can't say what would happen for any one individual. That's the downside to thinking about social policy and the defaults upstream in our broader environment. We can draw conclusions only about average effects—across groups of people.

But consider what happened to some survivors of Hurricane Katrina, the storm that devastated New Orleans in August 2005.[6] Researchers traced where 280 of the evacuees were relocated. These were mostly young women with children. None had much say about where they ended up. Their destinations were determined by random events such as evacuation traffic congestion and overcrowding in the shelters in nearby cities. Because these people could not select where to live, we can see whether they were influenced by their local environments, independent of their preferences for exercise and walking.

Most evacuees were moved from New Orleans to less urban, more sprawling communities with lower population density and less street accessibility. When they were contacted seven to nineteen months later, their weight had increased by an average of 5 percent. They weighed about nine pounds more! However, a few evacuees were moved to places just as dense and walkable as New Orleans. Their weight gain was essentially zero.

This study is important because it isolates one influence on our health and fitness. The magic ingredient in this study was whether the neighborhood had more opportunities to walk. This is largely a legislative issue—has your town built sidewalks that make it easy to get to the store and do everyday chores by walking? No question, having sidewalks outside is not the same as spending an hour at the gym. But a walkable community makes it possible to exercise even on days when you don't get to the gym—and even for people who never work out. It sets the default forces in our environment to favor the healthy option.

Our health is also tied to the transit we take, especially how we get to work. One study tracked about four thousand British commuters over two years to evaluate the effects of changing transit.[7] Some car commuters switched to more active transport, using train, bus, bike, or walking. In so doing, they reduced their body mass index

(BMI) by 0.32 points on average (about two pounds). Commuting distance also mattered. Long commutes of more than thirty minutes reduced BMI 2.25 points on average (about fourteen pounds). Active commuters at the beginning of the study who switched to driving their cars gained on average 0.34 points of BMI (about two pounds). We don't know why these folks changed their commuting habits. They might have moved closer to or farther from a transit line or started a new job. The most obese were probably the least likely to switch to active transport. But that's not the point. On average, commuters gained weight when they began driving; on average, they lost weight when they switched to mass transit, biking, or walking.

The question then becomes, "Would people in fact adopt more active transit if these options became more available?" Driving is the easy, affordable, familiar option for most Americans. Cars are so prevalent, it's difficult for many of us to imagine getting around any other way.

In Santa Monica, California, where I live, about half of all residents' trips are short-range, less than three miles. In 2017, to handle the congestion from short-range trips, electric Bird and Lime brand scooters became available for rent, like a bike-share program. There's a mobile app indicating where the closest one is located, they're easily reserved, and the cost in 2018 was just a dollar per trip plus fifteen cents per minute. The idea, according to Francie Stefan, Santa Monica's mobility manager, is to make the transportation ecosystem as varied as the biological one. The predominance of cars in the United States is like overplanting a single species. She says she wants "to offer a diversity of options that coexist for the long term."[8] But there are still kinks to work out, especially with safety. It's not clear if riders need helmets, and scooters on sidewalks have hit pedestrians. Once used, scooters are sometimes left to clutter city sidewalks and driveways.

Other cities, other forms of active transit: Portland, Washington, D.C., Minneapolis, Chicago, San Francisco, and Philadelphia are all building more bike lanes. The number of bike riders in these cities shows a corresponding increase in recent years.[9] New York City now has more than a thousand total miles of bike lanes, and cyclist commutes to work increased 80 percent between 2011 and 2016.[10] Minneapolis offers a fifty-one-mile ring of bike-only freeways, the Grand Rounds Scenic Byway. Despite Minnesota's notorious weather, the lanes have encouraged 5 percent of the city's residents to bike to work.

Social changes like these usually happen slowly at first and then can accelerate rapidly. Seat belts in cars are a prime example. Only a few people used them in the 1960s when U.S. car manufacturers started installing retractable seat belts, first as a purchase option and then as a legal requirement. The slow uptake of many innovations is due in part to old habits popping into mind before we have a chance to consider alternatives. Most of us don't even notice innovations until they have taken over. Seat belt use started to surge in the 1980s, as seat belt use laws were enacted by states. All U.S. cars now have seat belt detector reminders, and about 90 percent of drivers use belts. As we saw in chapter 14, people come to like the actions they repeat, and support for seat belt use grew rapidly as the belts were required by more states. Our concerns about safety and that habit of buckling up keep us using seat belts, probably even when driving through New Hampshire, which has no seat belt law for adults. Just acting, as we saw in the last chapter, can promote change in social opinions.

*

Realtors are recognizing the importance of default options in our neighborhoods (along with the human desire for self-sorting). From websites with community statistics (neighborhoodscout.com, niche.com), you can figure out what your life would probably be like in a new neighborhood. Your best guess is a combination of what you do now and the lifestyle of most people already living there.

Some of our most apparently ingrained habits are subject to these *map effects*. A study tracked more than six thousand Americans across eight years to see how their reported drinking habits responded to changes in the number of liquor stores nearby.[11] When density of stores increased, so did drinking. With each additional four stores per square mile, men increased their weekly consumption of beer by 32 percent. Women increased their wine intake by 16 percent.

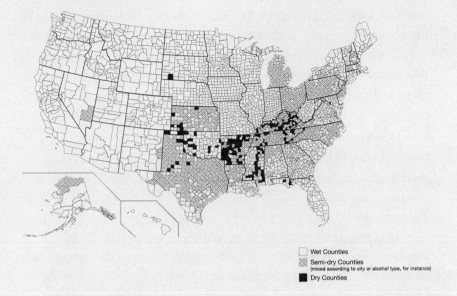

Wet Counties
Semi-dry Counties
(mixed according to city or alcohol type, for instance)
Dry Counties

Liquor laws vary greatly across the United States. In ten states, counties can prohibit sales of alcohol, as shown on the map above.[12] Counties shaded white are wet, meaning they sell alcohol, counties shaded black are dry, and counties shaded gray are partially dry, with some alcohol restrictions. Even within the white-shaded areas, options vary. New York City is wet, but neighborhoods differ, with some having as few as 5 alcohol outlets per square mile, and others as many as 132. Researchers in one study called residents of these areas to determine their binge drinking habits.[13] (A woman binge drinks if she consumes four or more drinks in two hours; a man, five or more.) In neighborhoods with 130 alcohol outlets, 13 percent of people surveyed drank heavily like this once a month or more. In neighborhoods with 20 outlets per square mile, only 8 percent were binge drinkers.

It's easy to understand how simple availability could have these effects. If you live in a dry area, you have to spend time and energy to get a drink. Not so easy to binge on impulse. The limited supply of alcohol nearby might also mean higher prices, increasing further the friction on drinking.

Of course, alcohol availability is probably something you didn't think much about the last time you moved. Even if you did, you may not have had much choice in the place you live. Often, we choose to live places for personal reasons that have nothing to do with the broader environmental forces that influence all of us. That's when the tools in this book become especially important, so that you can change your own habits to better meet your goals. That's also when civic participation becomes important. In a democracy, we can each speak out and vote to change upstream environments so that they become places in which the default option is the beneficial one for most of us.

*

Default options that affect all of us also become obvious over time. "Those who cannot remember the past are condemned to repeat it" is an often-repeated saying from philosopher George Santayana. This may be nowhere truer than in how much we eat.

American farm policy changed in the 1970s, about the time that the obesity epidemic started. After suffering a historic surge in basic commodity prices, citizens protested the inflated cost of basic foods, and the government changed the system of agricultural supports in ways that encourage overproduction. The policy changes were a political success—the price of food has not been a political issue since then. But the changes created a health risk. Since the mid-1970s, farmers have been incentivized to produce an additional 500 daily calories per person.[14] Two hundred calories have ended up on our plates, whereas the rest are used in other ways. The food industry grew, and so did we.

Meal sizes have gotten larger. According to the National Institutes of Health, portion sizes in restaurants have doubled or tripled in the past twenty years.[15] A standard-size bagel used to be three inches in diameter and 140 calories, but now is around six inches and 350 calories. A serving of spaghetti was one cup with sauce and three small meatballs, about 500 calories. Today? It's two cups of pasta with sauce and three large meatballs, totaling more than 1,000 calories. A turkey sandwich used to be 320 calories but now clocks in at 820. Portion distortion is obvious with fast food. This 2012 graphic by the U.S. Centers for Disease Control shows how much the size of an average fast-food meal has soared over time. Since the 1950s, sides of fries have tripled in size. Burgers quadrupled. Sodas have increased sixfold.

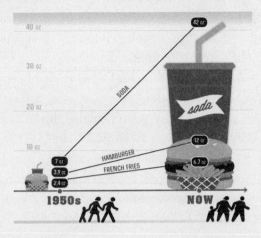

Restaurants serve, and we eat. As we saw in chapter 4, portion sizes matter in just the same way that the availability of alcohol in our neighborhoods matters. When it's easier to eat more, when portion and package sizes are larger, we simply do it. After all, it's already on our plate.[16] And once we start eating more, we start liking to eat more, and our biology further adjusts so that greater consumption becomes the norm.

Portion sizing is insidious, but each of us can make it easier to eat less by selecting which restaurants to frequent and whether to eat at home. We can choose what size packages of food to purchase at the store. Upstream defaults work on the average. They don't have to be part of the forces in your own personal environment.

A more policy-oriented solution, and one that has a harder edge than nudges, is taxation designed to limit consumption of empty-calorie foods. Two initiatives levied taxes on sugar-sweetened beverages: a tax of one cent per ounce in 2015 on such beverages in Berkeley, California, and a tax of one peso per liter in 2014 in Mexico. Both of these are large enough to be noticed by consumers at the cash register.

Taxes are always controversial in the United States. People prefer gentler nudges that preserve our sense of choice. But taxes have one advantage here. They not only add friction to the prior bad habit but also signal a new environmental shift. That shift is: we as a community no longer find this behavior virtuous. Collectively, the community is trying to diminish it. We're social creatures, and we read such cues. When the social standards change, people tend to get on board.

Soda manufacturers argued that customers would find their calories elsewhere. According to William Dermody, a media officer of the American Beverage Association, the trade group for soda makers, "Taxes and bans and restrictions don't change the behaviors that lead to obesity."[17] But as we know, taxes worked well for tobacco control. Smoking was cut in half by taxes on cigarettes, bans on smoking in public places, and restrictions on advertising.

In Berkeley, sales of sugar-sweetened beverages dropped 10 percent the year after the tax.[18] Residents simply bought other drinks. Sales of untaxed beverages increased by 4 percent. Water sales in particular increased by 16 percent.

In Mexico, before taxation, sugar-sweetened beverages accounted for 10 percent of daily calories.[19] Soda was one popular item. In the two years following the tax, consumption dropped about 8 percent. Mexicans bought other drinks instead. Sales of untaxed beverages increased 2 percent. As you'd expect, the poorest people were most hit by the tax, reducing their consumption 12 percent, whereas wealthier folks reduced by only 5 percent. In 2014, Mexico also imposed an 8 percent "nonessential energy-dense foods tax." This also worked. Purchase of these junk foods dropped by 6 percent in the first two years.[20]

The effects of these taxes on our weight have yet to be seen. Will taxes reduce obesity as they did smoking-related diseases? We know at this point that taxes decrease sales. That is an initial answer, but we have yet to see the health impact.

*

The impact of defaults is nowhere more evident than in our attempts to live a low-footprint, sustainable lifestyle. You may want to recycle and not contribute to overflowing landfills. If you live in a city, this is just easier. In 2016, 70 percent of U.S. city dwellers but only about

40 percent of rural residents had curbside recycling pickup. In addition, state of residence makes a difference.[21] In 2011, California, Maine, and Washington State recovered about 50 percent of their municipal solid waste. Oklahoma, Alaska, and Mississippi each recovered less than 5 percent. In those states, you have to work hard to find recycling bins and centers, and there's little curbside pickup to make it easy. For those residents, recycling takes real dedication.

What about energy use? More than half of all U.S. households already have smart meters installed that keep track.[22] Do you know if your house is one of these? Saving energy could be easy. However, of the 75 million government-installed smart meters in 2016, only a few thousand had in-home displays.[23] To access the meter, we'd need to log in to the energy company's website. Even worse, these websites don't give us feedback in real time. The information is there, we just don't have access to it.

Meters were a great policy idea because their feedback can put external forces to work reducing our energy use. With an in-home display, ignorance of how much energy we consume at home is no longer the default. More important than simply offering information, this device gives immediate rewards for turning off appliances. It takes only a few tries of turning down your air conditioner to figure out how much energy—and money—you save by setting the thermostat at 75 instead of 70 on a hot day.

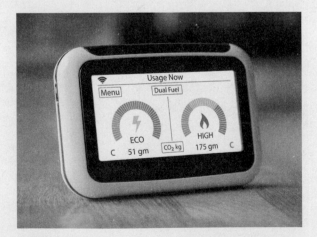

An experiment with more than four hundred Connecticut households showed just this.[24] During two summer months in 2011, residential electricity consumers were given in-home meter displays that showed real-time use, electricity price, estimated monthly use, and bill to date, much like the one on the previous page. They could watch the meter whir and the charges pile up. Or they could take action and reduce consumption. They were also sent text messages a day ahead to alert them about energy price increases during the hottest periods. A control group got only the messages, without smart meter displays. Across the two months of the study, the control group with just the notices reduced energy use during the most expensive, peak times by 7 percent. Those with smart meters were more successful. They lowered their energy use by 22 percent. The authors concluded that the meters reduced residential electricity greenhouse gases by 1–2 percent.

The meter is a textbook opportunity to form energy-saving habits. Your response of leaving the lights on gets immediate feedback in monetary charges. Your response of turning them off gets an immediate reward in that the charges drop. Repeat the rewarding activity over and over, and you have formed a habit of energy conservation (check meter, turn off electricity). It's a policy change that offers a built-in mechanism for habit formation. Now we just have to figure out how to get a display into homes in the United States.

Epilogue

This book is about something we all do every day, and for much of it. A huge part of our lives flows through our habit selves. This part of us moves more slowly than our conscious understanding. It takes a while to get going—but then it's pretty resilient. This part of ourselves is like a powerful, reliable laborer: always going, always on call. But it's occupied with what's right in front of it. That means there's still a place for "you"—for the you who's reading this book, the you who wants to lose some weight, or save some money, or get more out of your working day. That "you" has to set the goals. Then you have to use the habit-formation tools in this book to arrange your life into smart contexts full of the right driving forces, friction, and rewards.

Your new habit-charged life is going to be better for two reasons. One: you're going to get more done.

But the second one is just as important, and I've tried to emphasize it, too, throughout the book: it's a simpler, more integrated way to live your life.

We all live habitually already. Most of us just aren't aware of it. And because of that, we're ignoring a big part of who we are and why we do what we do. We're also ignoring all the many ways we could be doing things better.

It's exhausting and fruitless to live by motivation and willpower alone. You'll only fail yourself, time and again. You'll have all your goals and all your intentions, and you'll watch them get higher and higher and more out of reach. Your ideal life and your actual life will start to diverge more and more, and you'll feel that distance to be an indictment of your weakness and smallness of character.

It's a lie.

Living with your habit self will allow you to realize how much of you works in ways unbidden by your surface-level impulses and wishes. You contain depths. You can put those depths to work for you.

The principles you have learned make it easier to identify habits in all their varied forms. Bad habits then stop being impossible sink-holes in our lives and start being tractable challenges, ready to be faced and solved. Your good habits will no longer be inherent emanations of some mysterious essential character, and will become recognizable for what they are. Even better, they'll start to look like patterns on which you can build other, newer, better habits.

More than this, understanding habits normalizes the trials of changing behavior. The distance between repeated failure and endur-ing, successful change is not marked by personal fortitude or deter-mination. It is not a referendum on your personal worth. Even with amazing stamina, you can still fail. Instead, you cover that distance through simple steps, such as organizing contexts around you to en-courage enjoyable actions that meet your long-term goals. This is what those markedly successful people with seemingly high "self-control" already do. Like them, you can engage the help of supportive context cues. You can repeat actions sufficiently to form new habits that become the familiar norm. You can form habits that continue even when you are no longer experiencing the rewards.

This is the promise of a habit life well lived.

How to Stop Looking at Your Phone So Often

A Useful Story

You are likely someone who checks your work email after the conventional workday is done. I know this because in the latest Gallup data, 59 percent of U.S. workers who had work email accounts did so.[1] The downsides are clear, despite the advantages of flexible workplace scheduling. Greater electronic contact outside of work is tied to more stress, emotional exhaustion, and conflict between home and work.[2] This is not just people with bad jobs getting stressed. It seems to be specifically linked to being off work but still on call, accessible to supervisors, coworkers, or customers. In jobs that required extended work availability some days but not others, employees felt more anxious and less energetic and had worse moods on extended-work mornings, and even showed heightened spikes in cortisol, a stress-related hormone.[3]

Even if you are one of those lucky workers who never get a "Quick question" email from your boss at nine on a Wednesday evening, you're almost certainly still using your phone just as much as everyone else. You're just checking Facebook or Twitter or playing the latest game. Just hearing those pings of an incoming call or message attracts our attention and impairs performance at what we are currently doing.[4] And then there's the consequences for your relationship. *Cell phone snubbing* (or *partner phubbing*) is a new indicator of troubled relationships, as romantic partners check their phones instead of communicating with each other.[5] The predictable result is increased conflict and lessened intimacy in the future.[6]

There's no mystery as to why we and our relationships suffer. Getting sucked into your phone is like voluntarily putting on a set of horse blinders. You miss everything else in the world and just see the thing right in front of you—which is your phone. You check it when it vibrates, when it pings, when you sit down to breakfast, get out of your car, get into the office, get in the elevator, get out of the elevator . . . Because this is one of the most common habits in the world, I'm going to use it to illustrate the tools in this book.

The very first thing is to notice that you're using your phone too much in the first place. This perhaps sounds obvious, but remember that effective habits are effective precisely because they hide their workings from your consciousness. So you have to break through that. We've discussed some ways that it's possible to induce some of what you might call heightened habit awareness, such as taking advantage of the holistic disruption of big life events, but this is something that you need to come to grips with yourself—or, as likely as not, a friend or partner or coworker will. "Hey, you're obsessed—why don't you put that thing away for a bit?"

The next thing—and this is where our tools really kick in—is to *control the context cues* that activate and enable phone use. The game here is simple: remove the cues that make you grab the phone. The most straightforward way to do this is to leave your phone behind. Don't bring it when you sit down to breakfast or when you take a break at work for coffee and a doughnut (we'll work on that doughnut habit later). It'll be hard that first time, but unless you're an EMT, no one's going to notice that you're unavailable for fifteen minutes.

Maybe you grab the same three items every time you leave the house: your keys, your wallet/purse, and your phone? Well, only two of those three things are actually required for most places you want to go and most things you want to do. That leaving-the-house routine is a sticky cue for most of us. We like to be girded up for the outside world, ready to meet all eventualities. But try to think back all the way to 2004. Did your leaving-the-house routine include a trio of objects? Or was it just your keys and your wallet/purse? Did you survive?

Perhaps leaving your phone behind sounds like too obvious a method to get away from its grip. Fortunately, cell phones give us a plethora of cues to reengineer. You can readily *add friction to make phone use more difficult.* Silence it. Turn it off. Switch on your phone's Do Not Disturb mode so that only dedicated callers can get through to you. Removing the alerts removes cues to use and stops activating that unwanted thought, "Check phone."

There's more that you can do. Move your phone to a zippered pocket, say in your backpack, briefcase, or purse. Then you have to unzip it and reach down in there to retrieve it. Or you can turn it off after each use, so that each time, you have to go through the motions of powering it up again. This small delay doesn't seem like much to your conscious mind, but it adds friction and perhaps some frustration as well (really, the sensor didn't read my fingerprint or recognize my face *again?*). An easy way to add more delay and friction to your phone habit is simply to delete the Facebook app or your email app. At the very least, that'll mean you'll have to open the web browser and manually punch in "gmail.com" or "facebook.com" instead of relying on those companies' purposefully frictionless app designs.

Another way to add cost to checking is to *stack* a new, healthy action onto your existing cell phone habit. Even after tamping down the frequency, you're still going to look at it. So use this resilient (and likely necessary) habit to build another habit, one that's your own choice and oriented toward your own goals. What about if every time you check your phone, you call one member of your family just to say hello and have a quick, insignificant chat? One of those calls that feels great to get, a call for no reason. You are especially likely to delight older family members. And you'll be able to bolster some of those connections you've let attenuate (ironically, from too much activity on social media). If you really hold yourself to this new habit, you'll reconsider taking out your phone in the first place. Sometimes you just really don't want to talk to anyone. This raises the stakes of randomly checking your phone.

Whatever you choose to do to make using the phone more costly, do it consistently. With repetition, the initially difficult change becomes automated. The new action starts to be the one that automatically pops into mind, while the costs to your old habit persist.

Along with disrupting established cues and imposing friction, you can *make other actions easier*. Instead of looking at your phone, is there something else that you could quickly do instead? There's one viable alternative that I've personally witnessed many times: get a watch. How many times do you pull out your phone just to check the time or date, and you open up Facebook just because it's there in your hand . . . and you check your email because you see that you have a couple of new ones . . . and so it goes.

Instead of reaching into your pocket, lift up your wrist. Get a watch that you like and that you want to show off. Get a colorful watch, a calculator watch, a watch with a timer, an old mechanical watch (just don't get a smartwatch—that's cheating). This replacement behavior will immediately cut down on moments when you can get sucked into your phone.

And finally, *make not checking rewarding*. I can think of a very good reward for not checking your phone so often.

Let's say you sit down for a moment in a coffee shop. It's the middle of the afternoon. You're taking a quick break from the office. Of course, the moment is perfect to pull out your phone and check the latest news. But you've turned it off, put it in a zippered pocket, and would have to make a phone call to your aunt if you did use it. So you've successfully disrupted the cues and imposed restraining forces.

But there's no benefit in leaving yourself sitting there jonesing for your phone. Now give yourself something great to do. Give yourself something that's been stimulating people for centuries, something that's perfect for occupying your mind for a few minutes. Better than just occupying—something that'll expand your mind a bit, and fill in some blanks. Something that'll be useful for later that night at dinner by furnishing you with an interesting story or point to raise with your family. Something portable and durable. Something that'll nourish your whole self.

Got a good book to read?

Notes

1. Persistence and Change

1. Dan Ariely and Klaus Wertenbroch, "Procrastination, Deadlines, and Performance: Self-control by Precommitment," *Psychological Science* 13, no. 3 (2002): 219–24, doi:10.1111/1467-9280.00441; Janet Schwartz et al., "Healthier by Precommitment," *Psychological Science* 25, no. 2 (2014): 538–46, doi:10.1177/095 6797613510950.
2. "The ASMBS and NORC Survey on Obesity in America," NORC at the University of Chicago, accessed March 10, 2018, http://www.norc.org/Research /Projects/Pages/the-asmbsnorc-obesity-poll.aspx.
3. "New Insights into Americans' Perceptions and Misperceptions of Obesity Treatments, and the Struggles Many Face," NORC at the University of Chicago, October 2016, http://www.norc.org/PDFs/ASMBS%20Obesity/ASMBS %20NORC%20Obesity%20Poll_Brief%20B%20REV010917.pdf.
4. Icek Ajzen, "Residual Effects of Past on Later Behavior: Habituation and Reasoned Action Perspectives," *Personality and Social Psychology Review* 6, no. 2 (2002): 107–22, doi:10.1207/S15327957PSPR0602_02.
5. Rena R. Wing and Suzanne Phelan, "Long-term Weight Loss Maintenance," *The American Journal of Clinical Nutrition* 82, no. 1 (2005): 222S–25S, doi: 10.1093/ajcn/82.1.222S.
6. Wing and Phelan.
7. Interview with David Kirchhoff, former president and CEO of Weight Watchers, May 18, 2017.
8. David A. Kessler, *The End of Overeating: Taking Control of the Insatiable American Appetite* (Emmaus, PA: Rodale Books, 2009).
9. Daniel M. Wegner et al., "Paradoxical Effects of Thought Suppression," *Journal of Personality and Social Psychology* 53, no. 1 (1987): 5–14.
10. Daniel M. Wegner, "Ironic Processes of Mental Control," *Psychological Review* 101, no. 1 (1994): 34, doi:10.1037//0033-295x.101.1.34.

2. The Depths Beneath

1. Wendy Wood, Jeffrey M. Quinn, and Deborah A. Kashy, "Habits in Everyday Life: Thought, Emotion, and Action," *Journal of Personality and Social Psychology* 83, no. 6 (2002): 1281–97, doi:10.1037/0022-3514.83.6.1281.

2. Jeffrey M. Quinn and Wendy Wood, "Habits Across the Lifespan" (unpublished manuscript, Duke University, 2005).

3. Emily Pronin and Matthew B. Kugler, "People Believe They Have More Free Will Than Others," *Proceedings of the National Academy of Sciences* 107, no. 52 (2010): 22469–74, doi:10.1073/pnas.1012046108.

4. Richard E. Nisbett and Timothy D. Wilson, "Telling More Than We Can Know: Verbal Reports on Mental Processes," *Psychological Review* 84, no. 3 (1977): 231–59, doi:10.1037/0033-295X.84.3.231.

5. Nisbett and Wilson, 244.

6. Nisbett and Wilson, 244.

7. John H. Aldrich, Jacob M. Montgomery, and Wendy Wood, "Turnout as a Habit," *Political Behavior* 33, no. 4 (2011): 535–63, doi:10.1007/s11109-010-9148-3.

8. John T. Jost and David M. Amodio, "Political Ideology as Motivated Social Cognition: Behavioral and Neuroscientific Evidence," *Motivation and Emotion* 36, no. 1 (2012): 55–64, doi.10.1007/s11031-011-9260-7.

9. Partners Studio, "4 Reasons Why Over 50% Car Crashes Happen Closer to Home," *HuffPost*, December 14, 2017, https://www.huffingtonpost.co.za/2017/12/14/4-reasons-why-over-50-car-crashes-happen-closer-to-home_a_23307197.

10. "Odds of Dying," National Safety Council Injury Facts, 2016, https://injuryfacts.nsc.org/all-injuries/preventable-death-overview/odds-of-dying.

11. Kirsten Korosec, "2016 Was the Deadliest Year on American Roads in Nearly a Decade," *Fortune*, February 15, 2017, http://fortune.com/2017/02/15/traffic-deadliest-year/; *Global Status Report on Road Safety 2018* (Geneva: World Health Organization, 2018), https://www.who.int/violence_injury_prevention/road_safety_status/2018/en/.

12. Emily Gliklich, Rong Guo, and Regan W. Bergmark, "Texting While Driving: A Study of 1211 U.S. Adults with the Distracted Driving Survey," *Preventive Medicine Reports* 4 (2016): 486–89, doi:10.1016/j.pmedr.2016.09.003.

13. Brian J. Lucas and Loran F. Nordgren, "People Underestimate the Value of Persistence for Creative Performance," *Journal of Personality and Social Psychology* 109, no. 2 (2015): 232–43, doi:10.1037/pspa0000030.

3. Introducing Your Second Self

1. Edward C. Tolman, "Cognitive Maps in Rats and Men," *Psychological Review* 55, no. 4 (1948): 189–208, doi:10.1037/h0061626.

2. George A. Miller, "The Cognitive Revolution: A Historical Perspective," *Trends in Cognitive Sciences* 7, no. 3 (2003): 141–44, doi:10.1016/S1364-6613(03)00029-9.

3. George A. Miller, Eugene Galanter, and Karl H. Pribram, *Plans and the Structure of Behavior* (New York: Adams-Bannister-Cox, 1986), 2.

4. William James, *The Principles of Psychology,* vol. 1 (New York: Henry Holt, 1890; repr. Cosimo, 2007), 122.

5. Tara K. Patterson and Barbara J. Knowlton, "Subregional Specificity in Human Striatal Habit Learning: A Meta-Analytic Review of the fMRI Literature," *Current Opinion in Behavioral Sciences* 20 (2018): 75–82, doi:10.1016/j.cobeha .2017.10.005.

6. Richard M. Shiffrin and Walter Schneider, "Controlled and Automatic Human Information Processing: II. Perceptual Learning, Automatic Attending and a General Theory," *Psychological Review* 84, no. 2 (1977): 127–90, doi:10.1037 /0033-295X.84.2.127.

7. Walter Schneider and Richard M. Shiffrin, "Controlled and Automatic Human Information Processing: I. Detection, Search, and Attention," *Psychological Review* 84, no. 1 (1977): 1–66, doi:10.1037/0033-295X.84.1.1.

8. Christopher D. Adams and Anthony Dickinson, "Instrumental Responding Following Reinforcer Devaluation," *Quarterly Journal of Experimental Psychology 33B*, no. 2 (1981): 109–21, doi:10.1080/14640748108400816.

9. William James, *Habit* (New York: Henry Holt, 1890), 24.

10. David T. Neal, Wendy Wood, Jennifer S. Labrecque, and Phillippa Lally, "How Do Habits Guide Behavior? Perceived and Actual Triggers of Habits in Daily Life," *Journal of Experimental Social Psychology* 48, no. 2 (2012): 492–98, doi:10.1016/j.jesp.2011.10.011.

11. James, *Habit*, 24.

12. David E. Melnikoff and John A. Bargh, "The Mythical Number Two," *Trends in Cognitive Sciences* 22, no. 4 (2018): 280–93, doi:10.1016/j.tics.2018.02.001; David M. Amodio, "Social Cognition 2.0: An Interactive Memory Systems Account," *Trends in Cognitive Sciences* 23, no. 1 (2018): 21–33, doi:10.1016/j.tics .2018.10.002.

13. John A. Bargh, *Before You Know It: The Unconscious Reasons We Do What We Do* (New York: Touchstone, 2017).

14. John T. Wixted et al., "Initial Eyewitness Confidence Reliably Predicts Eyewitness Identification Accuracy," *American Psychologist* 70, no. 6 (2015): 515–26, doi:10.1037/a0039510.

15. Drake Baer, "The Scientific Reason Why Barack Obama and Mark Zuckerberg Wear the Same Outfit Every Day," *Business Insider*, April 28, 2015, http://www .businessinsider.com/barack-obama-mark-zuckerberg-wear-the-same-outfit -2015-4.

16. Alfred N. Whitehead, *An Introduction to Mathematics* (New York: Henry Holt, 1911).

17. Gary Klein, Roberta Calderwood, and Anne Clinton-Cirocco, "Rapid Decision Making on the Fire Ground: The Original Study Plus a Postscript," *Journal of Cognitive Engineering and Decision Making* 4, no. 3 (2010): 186–209, doi:10.15 18/155534310X12844000801203.

18. Klein et al., 193.

19. Klein et al., 194.

20. Interview with Clay Helton, head football coach at the University of Southern California, August 9, 2017, Los Angeles.

4. What About Knowledge?

1. Adwait Khare and J. Jeffrey Inman, "Habitual Behavior in American Eating Patterns: The Role of Meal Occasions," *Journal of Consumer Research* 32, no. 4 (2006): 567–75, doi:10.1086/500487.

2. Michael Mosley, "Five-A-Day Campaign: A Partial Success," BBC News, January 3, 2013, http://www.bbc.com/news/health-20858809.

3. Richard Doll and Richard Peto, "The Causes of Cancer: Quantitative Estimates of Avoidable Risks of Cancer in the United States Today," *JNCI: Journal of the National Cancer Institute* 66, no. 6 (1981): 1192–1308, doi:10.1093/jnci/66.6.1192.

4. Xia Wang et al., "Fruit and Vegetable Consumption and Mortality from All Causes, Cardiovascular Disease, and Cancer: Systematic Review and Dose-Response Meta-Analysis of Prospective Cohort Studies," *BMJ* 349 (2014): g4490, doi:10.1136/bmj.g4490.

5. Gloria Stables et al., "5 A Day Program Evaluation Research," in *5 A Day for Better Health Program Monograph*, eds. Gloria Stables and Jerianne Heimendinger (Rockville, MD: MasiMax, 2001), 89–111.

6. Sarah Stark Casagrande et al., "Have Americans Increased Their Fruit and Vegetable Intake? The Trends Between 1988 and 2002," *American Journal of Preventive Medicine* 32, no. 4 (2007): 257–63, doi:10.1016/j.amepre.2006.12.002.

7. Latetia V. Moore and Frances E. Thompson, "Adults Meeting Fruit and Vegetable Intake Recommendations—United States 2013," Centers for Disease Control and Prevention, *Morbidity and Mortality Weekly Report* 64, no. 26 (2015): 709–13, July 10, 2015, https://www.cdc.gov/mmwr/preview/mmwrhtml/mm6426a1.htm; NatCen Social Research, *Health Survey for England 2017* (London: NHS Digital, 2018), https://files.digital.nhs.uk/5B/B1297D/HSE%20report%20summary.pdf.

8. "What America Thinks: MetLife Foundation Alzheimer's Survey," MetLife Foundation, February 2011, https://www.metlife.com/content/dam/microsites/about/corporate-profile/alzheimers-2011.pdf.

9. Khare and Inman, "Habitual Behavior in American Eating Patterns."

10. Adwait Khare and J. Jeffrey Inman, "Daily, Week-Part, and Holiday Patterns in Consumers' Caloric Intake," *Journal of Public Policy and Marketing* 28, no. 2 (2009): 234–52, doi:10.1509/jppm.28.2.234.

11. Barbara J. Rolls, Liane S. Roe, and Jennifer S. Meengs, "The Effect of Large Portion Sizes on Energy Intake Is Sustained for 11 Days," *Obesity* 15, no. 6 (2007): 1535–43, doi:10.1038/oby.2007.182.

12. Pierre Chandon, "How Package Design and Packaged-Based Marketing Claims Lead to Overeating," *Applied Economic Perspectives and Policy* 35, no. 1 (2013): 7–31, doi:10.1093/aepp/pps028.

13. Nicole Diliberti et al., "Increased Portion Size Leads to Increased Energy Intake in a Restaurant Meal," *Obesity Research* 12, no. 3 (2004): 562–68, doi:10.1038/oby.2004.64.

14. Mindy F. Ji and Wendy Wood, "Purchase and Consumption Habits: Not Necessarily What You Intend," *Journal of Consumer Psychology* 17, no. 4 (2007): 261–76, doi:10.1016/S1057-7408(07)70037-2.

15. Barbara J. Knowlton and Tara K. Patterson, "Habit Formation and the Striatum," in *Behavioral Neuroscience of Learning and Memory*, eds. Robert E. Clark

and Stephen J. Martin, vol. 37 in *Current Topics in Behavioral Neurosciences* (Cham, Switzerland: Springer International, 2018), 275–95, doi:10.1007/7854 _2016_451.

16. Henry H. Yin and Barbara J. Knowlton, "The Role of the Basal Ganglia in Habit Formation," *Nature Reviews Neuroscience* 7, no. 6 (2006): 464–76, doi:10.1038/ nrn1919.

17. Bernard W. Balleine and John P. O'Doherty, "Human and Rodent Homologies in Action Control: Corticostriatal Determinants of Goal-Directed and Habitual Action," *Neuropsychopharmacology* 35, no. 1 (2010): 48–69, doi:10.1038/npp .2009.131.

18. Barbara J. Knowlton, Jennifer A. Mangels, and Larry R. Squire, "A Neostriatal Habit Learning System in Humans," *Science* 273, no. 5280 (1996): 1399–1402, doi:10.1126/science.273.5280.1399; Peter Redgrave et al., "Goal-Directed and Habitual Control in the Basal Ganglia: Implications for Parkinson's Disease," *Nature Reviews Neuroscience* 11, no. 11 (2010): 760–72, doi:10.1038/ nrn2915.

19. Knowlton and Patterson, "Habit Formation and the Striatum"; Tara K. Patterson and Barbara J. Knowlton, "Subregional Specificity in Human Striatal Habit Learning: A Meta-Analytic Review of the fMRI Literature," *Current Opinion in Behavioral Sciences* 20 (2018): 75–82, doi:10.1016/j.cobeha.2017 .10.005.

20. Guy Itzchakov, Liad Uziel, and Wendy Wood, "When Attitudes and Habits Don't Correspond: Self-Control Depletion Increases Persuasion but Not Behavior," *Journal of Experimental Social Psychology* 75 (2018): 1–10, doi:10.1016/j .jesp.2017.10.011.

21. A. N. Whitehead, *An Introduction to Mathematics* (New York: Henry Holt, 1911).

22. Jonathan St. B. T. Evans and Keith E. Stanovich, "Dual-Process Theories of Higher Cognition: Advancing the Debate," *Perspectives on Psychological Science* 8, no. 3 (2013): 223–41, doi:10.1177/1745691612460685.

23. Amitai Shenhav et al., "Toward a Rational and Mechanistic Account of Mental Effort," *Annual Review of Neuroscience* 40 (2017): 99–124, doi:10.1146/annu rev-neuro-072116-031526.

24. Shenhav et al.

5. What About Self-Control?

1. Walter Mischel and Ebbe B. Ebbesen, "Attention in Delay of Gratification," *Journal of Personality and Social Psychology* 16, no. 2 (1970): 329–37, doi:10.1037/ h0029815.

2. Yuichi Shoda, Walter Mischel, and Philip K. Peake, "Predicting Adolescent Cognitive and Self-Regulatory Competencies from Preschool Delay of Gratification: Identifying Diagnostic Conditions," *Developmental Psychology* 26, no. 6 (1990): 978–86, doi:10.1037/0012-1649.26.6.978.

3. Tanya R. Schlam et al., "Preschoolers' Delay of Gratification Predicts Their Body Mass 30 Years Later," *The Journal of Pediatrics* 162, no. 1 (2013): 90–93, doi:10.1016/j.jpeds.2012.06.049.

4. Shoda, Mischel, and Peake, "Predicting Adolescent Cognitive and Self-Regulatory Competencies."

5. Jeffrey M. Quinn et al., "Can't Control Yourself? Monitor Those Bad Habits," *Personality and Social Psychology Bulletin* 36, no. 4 (2010): 499–511, doi: 10.1177/0146167209360665.

6. June P. Tangney, Roy F. Baumeister, and Angie Luzio Boone, "High Self-Control Predicts Good Adjustment, Less Pathology, Better Grades, and Interpersonal Success," *Journal of Personality* 72, no. 2 (2004): 274, doi:10.1111/j.0022-3506.2004.00263.x.

7. Tangney, Baumeister, and Boone.

8. Eli J. Finkel and W. Keith Campbell, "Self-Control and Accommodation in Close Relationships: An Interdependence Analysis," *Journal of Personality and Social Psychology* 81, no. 2 (2001): 263–77, doi:10.1037//0022-3514.81.2.263.

9. Kirby Deater-Deckard et al., "Maternal Working Memory and Reactive Negativity in Parenting," *Psychological Science* 21, no. 1 (2010): 75–79, doi:10.1177/0956797609354073.

10. Camilla Strömbäck et al., "Does Self-Control Predict Financial Behavior and Financial Well-Being?" *Journal of Behavioral and Experimental Finance* 14 (2017): 30–38, doi:10.1016/j.jbef.2017.04.002.

11. Carmen Keller, Christina Hartmann, and Michael Siegrist, "The Association between Dispositional Self-Control and Longitudinal Changes in Eating Behaviors, Diet Quality, and BMI," *Psychology and Health* 31, no. 11 (2016): 1311–27, doi:10.1080/08870446.2016.1204451.

12. Wilhelm Hofmann et al., "Everyday Temptations: An Experience Sampling Study of Desire, Conflict, and Self-Control," *Journal of Personality and Social Psychology* 102, no. 6 (2012): 1318–35, doi:10.1037/a0026545.

13. Brian M. Galla and Angela L. Duckworth, "More Than Resisting Temptation: Beneficial Habits Mediate the Relationship between Self-Control and Positive Life Outcomes," *Journal of Personality and Social Psychology* 109, no. 3 (2015): 508–25, doi:10.1037/pspp0000026.

14. Galla and Duckworth.

15. Galla and Duckworth.

16. Denise T. D. de Ridder et al., "Taking Stock of Self-Control: A Meta-Analysis of How Trait Self-Control Relates to a Wide Range of Behaviors," *Personality and Social Psychology Review* 16, no. 1 (2012): 76–99, doi:10.1177/1088868311418749.

17. De Ridder et al., 91.

18. Ruth Umoh, "Bill Gates Said He Had to Quit This Common Bad Habit Before He Became Successful," CNBC, March 16, 2018, https://www.cnbc.com/2018/03/16/bill-gates-quit-this-bad-habit-before-he-became-successful.html.

19. "I'm Bill Gates, Co-Chair of the Bill and Melinda Gates Foundation. Ask Me Anything," Reddit, accessed May 14, 2018, https://www.reddit.com/r/IAmA/comments/49jkhn/im_bill_gates_cochair_of_the_bill_melinda_gates.

20. Umoh, "Bill Gates Said He Had to Quit This Common Bad Habit."

21. Bill Gates, *Business @ the Speed of Thought: Succeeding in the Digital Economy* (New York: Hachette Book Group, 1999).

22. Christian Crandall and Monica Biernat, "The Ideology of Anti-Fat Attitudes," *Journal of Applied Social Psychology* 20, no. 3 (1990): 227–43, doi:10.1111/j.1559-1816.1990.tb00408.x.

23. Pei-Ying Lin, Wendy Wood, and John Monterosso, "Healthy Eating Habits Protect against Temptations," *Appetite* 103 (2016): 432–40, doi:10.1016/j.appet.2015.11.011.

24. Angela L. Duckworth, Tamar Szabó Gendler, and James J. Gross, "Situational Strategies for Self-Control," *Perspectives on Psychological Science* 11, no. 1 (2016): 35–55, doi:10.1177/1745691615623247.

6. Context

1. Lydia Saad, "U.S. Smoking Rate Still Coming Down," Gallup, July 24, 2008, https://news.gallup.com/poll/109048/us-smoking-rate-still-coming-down.aspx.

2. "Tobacco-Related Mortality," Centers for Disease Control and Prevention, May 15, 2017, https://www.cdc.gov/tobacco/data_statistics/fact_sheets/health_effects/tobacco_related_mortality/index.htm.

3. Lydia Saad, "Tobacco and Smoking," Gallup, August 15, 2002, http://www.gallup.com/poll/9910/tobacco-smoking.aspx.

4. *Smoking and Health: A Report of the Surgeon General: Appendix: Cigarette Smoking in the United States, 1950–1978* (United States Public Health Service, Office on Smoking and Health, 1979), https://profiles.nlm.nih.gov/ps/access/nnbcph.pdf.

5. "Burden of Tobacco Use in the U.S.," Centers for Disease Control and Prevention, last modified April 23, 2018, https://www.cdc.gov/tobacco/campaign/tips/resources/data/cigarette-smoking-in-united-states.html; "Tobacco Data and Statistics," World Health Organization, accessed February 16, 2019, http://www.euro.who.int/en/health-topics/disease-prevention/tobacco/data-and-statistics.

6. "Cigarette Smoking and Tobacco Use Among People of Low Socioeconomic Status," Centers for Disease Control and Prevention, last modified August 21, 2018, https://www.cdc.gov/tobacco/disparities/low-ses/index.htm.

7. U.S. Department of Health and Human Services, *The Health Consequences of Smoking: 50 Years of Progress. A Report of the Surgeon General* (Atlanta, GA: U.S. Department of Health and Human Services, Centers for Disease Control and Prevention, National Center for Chronic Disease Prevention and Health Promotion, Office on Smoking and Health, 2014), 868.

8. "Quitting Smoking Among Adults—United States, 2000–2015," Centers for Disease Control and Prevention, January 6, 2017, https://www.cdc.gov/tobacco/data_statistics/mmwrs/byyear/2017/mm6552a1/highlights.htm.

9. Eleni Vangeli et al., "Predictors of Attempts to Stop Smoking and Their Success in Adult General Population Samples: A Systematic Review," *Addiction* 106, no. 12 (2011): 2110–21, doi:10.1111/j.1360-0443.2011.03565.x.

10. "Quitting Smoking Among Adults—United States, 2000–2015."

11. Michael Chaiton et al., "Estimating the Number of Quit Attempts It Takes to Quit Smoking Successfully in a Longitudinal Cohort of Smokers," *BMJ Open* 6, no. 6 (2016): e011045, doi:10.1136/bmjopen-2016-011045.

12. Jody Brumage, "The Public Health Cigarette Smoking Act of 1970," Robert C. Byrd Center, July 25, 2017, https://www.byrdcenter.org/byrd-center-blog/the-public-health-cigarette-smoking-act-of-1970.

13. "State and Local Comprehensive Smoke-Free Laws for Worksites, Restaurants, and Bars—United States, 2015," Centers for Disease Control and Prevention, last modified August 24, 2017, https://www.cdc.gov/mmwr/volumes/65/wr/mm6524a4.htm.

14. Emily M. Mader et al., "Update on Performance in Tobacco Control: A Longitudinal Analysis of the Impact of Tobacco Control Policy and the US Adult Smoking Rate, 2011–2013," *Journal of Public Health Management and Practice* 22, no. 5 (2016): E29–E35, doi:10.1097/phh.0000000000000358; Mader et al. also found that smoking cessation services did not have a significant impact on smoking rate, but, noting that other studies have found a positive impact, suggested more such services merit increased funding.

15. Justin McCarthy, "In U.S., Smoking Rate Lowest in Utah, Highest in Kentucky," Gallup, March 13, 2014, http://www.gallup.com/poll/167771/smoking-rate-lowest-utah-highest-kentucky.aspx.

16. Sheina Orbell and Bas Verplanken, "The Automatic Component of Habit in Health Behavior: Habit as Cue-Contingent Automaticity," *Health Psychology* 29, no. 4 (2010): 374–83, doi:10.1037/a0019596.

17. Morgan Scarboro, "How High Are Cigarette Taxes in Your State?" Tax Foundation, May 10, 2017, https://taxfoundation.org/state-cigarette-taxes/.

18. "Map of Excise Tax Rates on Cigarettes," Centers for Disease Control and Prevention, January 2, 2018, https://www.cdc.gov/statesystem/excisetax.html.

19. "Map of Current Cigarette Use Among Adults," Centers for Disease Control and Prevention, September 19, 2017, https://www.cdc.gov/statesystem/cigaretteuseadult.html.

20. Stanton A. Glantz, "Tobacco Taxes Are Not the Most Effective Tobacco Control Policy (As Actually Implemented)," UCSF Center for Tobacco Control Research and Education, January 11, 2014, https://tobacco.ucsf.edu/tobacco-taxes-are-not-most-effective-tobacco-control-policy-actually-implemented.

21. Thomas R. Kirchner et al., "Geospatial Exposure to Point-of-Sale Tobacco: Real-Time Craving and Smoking-Cessation Outcomes," *American Journal of Preventive Medicine* 45, no. 4 (2013): 379–85, doi:10.1016/j.amepre.2013.05.016; see also Steven J. Hoffman and Charlie Tan, "Overview of Systematic Reviews on the Health-Related Effects of Government Tobacco Control Policies," *BMC Public Health* 15 (2015): 744, doi:10.1186/s12889-015-2041-6; and Christopher P. Morley and Morgan A. Pratte, "State-Level Tobacco Control and Adult Smoking Rate in the United States: An Ecological Analysis of Structural Factors," *Journal of Public Health Management and Practice* 19, no. 6 (2013): E20–E27, doi:10.1097/PHH.0b013e31828000de.

22. Kurt Lewin, "Frontiers in Group Dynamics: Concept, Method and Reality in Social Science; Social Equilibria and Social Change," *Human Relations* 1, no. 1 (1947): 5–41, doi:10.1177/001872674700100103.

23. Interview with Professor M. Keith Chen, former head of economic research for Uber, May 15, 2017, Santa Monica, CA.

24. Gregory J. Privitera and Faris M. Zuraikat, "Proximity of Foods in a Competi-

tive Food Environment Influences Consumption of a Low Calorie and a High Calorie Food," *Appetite* 76 (2014): 175–79, doi:10.1016/j.appet.2014.02.004.

25. Valérie J. V. Broers et al., "A Systematic Review and Meta-Analysis of the Effectiveness of Nudging to Increase Fruit and Vegetable Choice," *European Journal of Public Health* 27, no. 5 (2017): 912–20, doi:10.1093/eurpub/ckx085; Tamara Bucher et al., "Nudging Consumers Towards Healthier Choices: A Systematic Review of Positional Influences on Food Choice," *British Journal of Nutrition* 115, no. 12 (2016): 2252–63, doi:10.1017/s0007114516001653.

26. Akihiko Michimi and Michael C. Wimberly, "Associations of Supermarket Accessibility with Obesity and Fruit and Vegetable Consumption in the Conterminous United States," *International Journal of Health Geographics* 9, no. 1 (2010): 49, doi:10.1186/1476-072x-9-49; Paul L. Robinson et al., "Does Distance Decay Modelling of Supermarket Accessibility Predict Fruit and Vegetable Intake by Individuals in a Large Metropolitan Area?" *Journal of Health Care for the Poor and Underserved* 24, no. 1A (2013): 172–85, doi:10.1353/hpu .2013.0049.

27. J. Nicholas Bodor et al., "Neighbourhood Fruit and Vegetable Availability and Consumption: The Role of Small Food Stores in an Urban Environment," *Public Health Nutrition* 11, no. 4 (2008): 413–20, doi:10.1017/s1368980007000493.

28. Alexandra E. Evans et al., "Introduction of Farm Stands in Low-Income Communities Increases Fruit and Vegetable among Community Residents," *Health and Place* 18, no. 5 (2012): 1137–43, doi:10.1016/j.healthplace.2012.04.007.

29. Rachel Bachman, "How Close Do You Need to Be to Your Gym?" *The Wall Street Journal*, March 21, 2017, https://www.wsj.com/articles/how-close-do-you -need-to-be-to-your-gym-1490111186.

30. Leon Festinger, Stanley Schachter, and Kurt Back, *Social Pressures in Informal Groups; A Study of Human Factors in Housing* (New York: Harper, 1950).

31. Erin Frey and Todd Rogers, "Persistence: How Treatment Effects Persist After Interventions Stop," *Policy Insights from the Behavioral and Brain Sciences* 1, no. 1 (2014): 172–79, doi:10.1177/2372732214550405.

32. Lenny R. Vartanian et al., "Modeling of Food Intake: A Meta-Analytic Review," *Social Influence* 10, no. 3 (2015): 119–36, doi:10.1080/15534510.2015.1 008037; Tegan Cruwys, Kirsten E. Bevelander, and Roel C. J. Hermans, "Social Modeling of Eating: A Review of When and Why Social Influence Affects Food Intake and Choice," *Appetite* 86 (2015): 3–18, doi:10.1016/j.appet.2014 .08.035.

33. Lenny R. Vartanian et al., "Conflicting Internal and External Eating Cues: Impact on Food Intake and Attributions," *Health Psychology* 36, no. 4 (2017): 365–69, doi:10.1037/hea0000447; Samantha Spanos et al., "Failure to Report Social Influences on Food Intake: Lack of Awareness or Motivated Denial?" *Health Psychology* 33, no. 12 (2014): 1487–94, doi:10.1037/hea0000008.

34. Scott E. Carrell, Mark Hoekstra, and James E. West, "Is Poor Fitness Contagious? Evidence from Randomly Assigned Friends," *Journal of Public Economics* 95, nos. 7–8 (2011): 657–63, www.nber.org/papers/w16518.

35. Derek J. Koehler, Rebecca J. White, and Leslie K. John, "Good Intentions, Optimistic Self-Predictions, and Missed Opportunities," *Social Psychological and Personality Science* 2, no. 1 (2011): 90–96, doi:10.1177/1948550610375722.

36. Lee D. Ross, Teresa M. Amabile, and Julia L. Steinmetz, "Social Roles, Social Control, and Biases in Social-Perception Processes," *Journal of Personality and Social Psychology* 35, no. 7 (1977): 485–94, doi:10.1037/0022-3514.35.7.485.

7. Repetition

1. Jayne A. Fulkerson et al., "Family Dinner Meal Frequency and Adolescent Development: Relationships with Developmental Assets and High-Risk Behaviors," *Journal of Adolescent Health* 39, no. 3 (2006): 337–45, doi:10.1016/j.jadohealth.2005.12.026; Amber J. Hammons and Barbara H. Fiese, "Is Frequency of Shared Family Meals Related to the Nutritional Health of Children and Adolescents?" *Pediatrics* 127, no. 6 (2011): E1565–74, doi:10.1542/peds.2010-1440.
2. Maxwell Maltz, *Psycho-Cybernetics* (New York: Pocket Books, 1989).
3. Phillippa Lally et al., "How Are Habits Formed: Modelling Habit Formation in the Real World," *European Journal of Social Psychology* 40, no. 6 (2010): 998–1009, doi:10.1002/ejsp.674.
4. Paschal Sheeran et al., "Paradoxical Effects of Experience: Past Behavior Both Strengthens and Weakens the Intention-Behavior Relationship," *Journal of the Association of Consumer Research* 2, no. 3 (2017): 309–18, doi:10.1086/691216.
5. Interview with Professor M. Keith Chen, former head of economic research for Uber, May 15, 2017, Santa Monica, CA.
6. Brian M. Galla and Angela L. Duckworth, "More Than Resisting Temptation: Beneficial Habits Mediate the Relationship between Self-Control and Positive Life Outcomes," *Journal of Personality and Social Psychology* 109, no. 3 (2015): 508–25, doi:10.1037/pspp0000026.
7. Unna N. Danner, Henk Aarts, and Nanne K. de Vries, "Habit vs. Intention in the Prediction of Future Behaviour: The Role of Frequency, Context Stability and Mental Accessibility of Past Behaviour," *British Journal of Social Psychology* 47, no. 2 (2008): 245–65, doi:10.1348/014466607x230876.
8. Bas Verplanken, Henk Aarts, and Ad van Knippenberg, "Habit, Information Acquisition, and the Process of Making Travel Mode Choices," *European Journal of Social Psychology* 27, no. 5 (1997): 539–60, doi:10.1002/(SICI)1099-0992(199709/10)27:5<539::AID-EJSP831>3.0.CO;2-A; Henk Aarts, Bas Verplanken, and Ad van Knippenberg, "Habit and Information Use in Travel Mode Choices," *Acta Psychologica* 96, no. 1–2 (1997): 1–14, doi:10.1016/s0001-6918(97)00008-5.
9. Steven S. Posavac, Frank R. Kardes, and J. Joško Brakus, "Focus Induced Tunnel Vision in Managerial Judgment and Decision Making: The Peril and the Antidote," *Organizational Behavior and Human Decision Processes* 113, no. 2 (2010): 102–11, doi:10.1016/j.obhdp.2010.07.002.
10. Christopher J. Armitage, "Can the Theory of Planned Behavior Predict the Maintenance of Physical Activity?" *Health Psychology* 24, no. 3 (2005): 235–45, doi:10.1037/0278-6133.24.3.235.
11. Will Durant, *The Story of Philosophy: The Lives and Opinions of the World's Greatest Philosophers* (New York: Pocket Books, 1926, 1954), 87.
12. Malcolm Gladwell, *Outliers: The Story of Success* (New York: Little, Brown, 2008).

13. Benjamin Morris, "Stephen Curry Is the Revolution," *FiveThirtyEight*, December 3, 2015, http://fivethirtyeight.com/features/stephen-curry-is-the-revolution.

14. Michael Rothman, "Stephen and Ayesha Curry: Inside Our Whirlwind Life," ABC News, accessed May 18, 2018, https://abcnews.go.com/Entertainment /fullpage/stephen-ayesha-curry-inside-whirlwind-life-34207323.

15. Mark J. Burns, "Success Is Not an Accident: What Sports Business Millennials Can Learn from NBA MVP Stephen Curry," *Forbes*, June 13, 2015, https:// www.forbes.com/sites/markjburns/2015/06/13/success-is-not-an-accident-what -sports-business-millennials-can-learn-from-nba-mvp-stephen-curry-2 /#62c34b3d15fb.

16. Brooke N. Macnamara, David Z. Hambrick, and Frederick L. Oswald, "Deliberate Practice and Performance in Music, Games, Sports, Education, and Professions: A Meta-Analysis," *Psychological Science* 25, no. 8 (2014): 1608–18, doi:10.1177 /0956797614535810.

8. Reward

1. Henry H. Yin and Barbara J. Knowlton, "The Role of the Basal Ganglia in Habit Formation," *Nature Reviews Neuroscience* 7, no. 6 (2006): 464–76, doi:10.1038/ nrn1919.

2. Wolfram Schultz, "Dopamine Reward Prediction Error Coding," *Dialogues in Clinical Neuroscience* 18, no. 1 (2016): 23–32.

3. Roy A. Wise, "Dopamine and Reward: The Anhedonia Hypothesis 30 Years On," *Neurotoxicity Research* 14, no. 2–3 (2008): 169–83, doi:10.1007/bf03033808; Wolfram Schultz, "Neuronal Reward and Decision Signals: From Theories to Data," *Physiological Reviews* 95, no. 3 (2015): 853–951, doi:10.1152/physrev .00023.2014.

4. Schultz, "Neuronal Reward and Decision Signals."

5. Diane R. Follingstad and Maryanne Edmundson, "Is Psychological Abuse Reciprocal in Intimate Relationships? Data from a National Sample of American Adults," *Journal of Family Violence* 25, no. 5 (2010): 495–508, doi:10.1007/ s10896-010-9311-y.

6. Wolfram Schultz, "Dopamine Reward Prediction-Error Signalling: A Two-Component Response," *Nature Reviews Neuroscience* 17, no. 3 (2016): 183–95, doi:10.1038/nrn.2015.26.

7. Tomomi Shindou et al., "A Silent Eligibility Trace Enables Dopamine-Dependent Synaptic Plasticity for Reinforcement Learning in the Mouse Striatum," *European Journal of Neuroscience* (2018): 1–11, doi:10.1111/ejn.13921.

8. Volkswagen, "The Fun Theory 1—Piano Staircase Initiative," October 26, 2009, video, 1:47, https://www.youtube.com/watch?v=SByymar3bds.

9. Volkswagen, "The Fun Theory 2—An Initiative of Volkswagen: The World's Deepest Bin," October 26, 2009, video, 1:26, https://www.youtube.com/watch ?v=qRgWttqFKu8.

10. Benjamin Gardner and Phillippa Lally, "Does Intrinsic Motivation Strengthen Physical Activity Habit? Modeling Relationships between Self-Determination, Past Behaviour, and Habit Strength," *Journal of Behavioral Medicine* 36, no. 5

(2013): 488–97, doi:10.1007/s10865-012-9442-0; for similar findings with eating fruits and vegetables, see Amelie U. Wiedemann et al., "Intrinsic Rewards, Fruit and Vegetable Consumption, and Habit Strength: A Three-Wave Study Testing the Associative-Cybernetic Model," *Applied Psychology: Health and Well-Being* 6, no. 1 (2014): 119–34, doi:10.1111/aphw.12020.

11. Pei-Ying Lin, Wendy Wood, and John Monterosso, "Healthy Eating Habits Protect against Temptations," *Appetite* 103 (2016): 432–40, doi:10.1016/j.appet.2015.11.011.

12. Eleni Mantzari et al., "Personal Financial Incentives for Changing Habitual Health-Related Behaviors: A Systematic Review and Meta-Analysis," *Preventive Medicine* 75 (2015): 75–85, doi:10.1016/j.ypmed.2015.03.001.

13. Jeffrey T. Kullgren et al., "Individual Versus Group-Based Financial Incentives for Weight Loss: A Randomized, Controlled Trial," *Annals of Internal Medicine* 158, no. 7 (2013): 505–14, doi:10.7326/0003-4819-158-7-201304020-00002.

14. Wendy Wood and David T. Neal, "Healthy through Habit: Interventions for Initiating and Maintaining Health Behavior Change," *Behavioral Science and Policy* 2, no. 1 (2016): 71–83, doi:10.1353/bsp.2016.0008.

15. Rebecca Greenfield, "Workplace Wellness Programs Really Don't Work," *Bloomberg*, January 26, 2018, https://www.bloomberg.com/news/articles/2018-01-26/workplace-wellness-programs-really-don-t-work.

16. John Rosengren, "How Casinos Enable Gambling Addicts," *The Atlantic*, December 2016, https://www.theatlantic.com/magazine/archive/2016/12/losing-it-all/505814/.

17. Patrick Anselme, "Dopamine, Motivation, and the Evolutionary Significance of Gambling-Like Behaviour," *Behavioural Brain Research* 256 (2013): 1–4, doi:10.1016/j.bbr.2013.07.039.

18. Lisa Eadicicco, "Americans Check Their Phones 8 Billion Times a Day," *Time*, December 15, 2015, http://time.com/4147614/smartphone-usage-us-2015.

19. Alicia L. DeRusso et al., "Instrumental Uncertainty as a Determinant of Behavior under Interval Schedules of Reinforcement," *Frontiers in Integrative Neuroscience* 4 (2010): 17, doi:10.3389/fnint.2010.00017.

20. Luxi Shen, Ayelet Fishbach, and Christopher K. Hsee, "The Motivating-Uncertainty Effect: Uncertainty Increases Resource Investment in the Process of Reward Pursuit," *Journal of Consumer Research* 41, no. 5 (2015): 1301–15, doi:10.1086/679418.

21. Kellie Ell, "Video Game Industry Is Booming with Continued Revenue," CNBC, July 18, 2018, https://www.cnbc.com/2018/07/18/video-game-industry-is-booming-with-continued-revenue.html.

22. Erol Ozcelik, Nergiz Ercil Cagiltay, and Nese Sahin Ozcelik, "The Effect of Uncertainty on Learning in Game-Like Environments," *Computers and Education* 67 (2013): 12–20, doi:10.1016/j.compedu.2013.02.009; see also Paul A. Howard-Jones et al., "Gamification of Learning Deactivates the Default Mode Network," *Frontiers in Psychology* 6 (2016): 1891, doi:10.3389/fpsyg.2015.01891.

23. Zakkoyya H. Lewis, Maria C. Swartz, and Elizabeth J. Lyons, "What's the Point? A Review of Reward Systems Implemented in Gamification Interventions," *Games for Health Journal* 5, no. 2 (2016): 93–99, doi:10.1089/g4h.2015.0078.

24. Yin and Knowlton, "The Role of the Basal Ganglia in Habit Formation."

25. Christopher D. Adams, "Variations in the Sensitivity of Instrumental Respond-ing to Reinforcer Devaluation," *The Quarterly Journal of Experimental Psychology Section B* 34, no. 2b (1982): 77–98, doi:10.1080/14640748208400878; Anthony Dickinson, "Actions and Habits: The Development of Behavioural Autonomy," *Philosophical Transactions of the Royal Society of London. B: Biological Sciences* 308, no. 1135 (1985): 67–78, doi:10.1098/rstb.1985.0010.

26. David T. Neal et al., "The Pull of the Past: When Do Habits Persist Despite Conflict with Motives?" *Personality and Social Psychology Bulletin* 37, no. 11 (2011): 1428–37, doi:10.1177/0146167211419863.

27. Justine Burns, Brendan Maughan-Brown, and Âurea Mouzinho, "Washing with Hope: Evidence from a Hand-Washing Pilot Study among Children in South Africa," *BMC Public Health* 18 (2018): 709, doi:10.1186/s12889-018-5573-8; Abigail Sellman, Justine Burns, and Brendan Maughan-Brown, "Hand-washing Behaviour and Habit Formation in the Household: Evidence of Spillovers from a Pilot Randomised Evaluation in South Africa," SALDRU Working Paper Series, no. 226 (2018).

28. David Neal et al., *The Science of Habit: Creating Disruptive and Sticky Behavior Change in Handwashing Behavior* (Washington, D.C.: USAID/WASHplus Proj-ect, 2015).

9. Consistency Is for Closers

1. Navin Kaushal and Ryan E. Rhodes, "Exercise Habit Formation in New Gym Members: A Longitudinal Study," *Journal of Behavioral Medicine* 38, no. 4 (2015): 652–63, doi:10.1007/s10865-015-9640-7.

2. L. Alison Phillips, Howard Leventhal, and Elaine A. Leventhal, "Assessing Theo-retical Predictors of Long-Term Medication Adherence: Patients' Treatment-Related Beliefs, Experiential Feedback and Habit Development," *Psychology and Health* 28, no. 10 (2013): 1135–51, doi:10.1080/08870446.2013.793798.

3. Gerard J. Molloy, Heather Graham, and Hannah McGuinness, "Adherence to the Oral Contraceptive Pill: A Cross-Sectional Survey of Modifiable Behav-ioural Determinants," *BMC Public Health* 12 (2012): 838, doi:10.1186/1471-2458-12-838.

4. Phillips, Leventhal, and Leventhal, "Assessing Theoretical Predictors of Long-Term Medication Adherence."

5. Ellen Berscheid and Hilary Ammazzalorso, "Emotional Experience in Close Relationships," in *Blackwell Handbook of Social Psychology: Interpersonal Pro-cesses,* eds. Garth Fletcher and Margaret Clark (Malden, MA: Blackwell Pub-lishers, 2001), 308–30; Ellen Berscheid and Pamela Regan, *The Psychology of Interpersonal Relationships* (New York: Pearson, 2005; repr. Routledge, 2016).

6. John G. Holmes and Susan D. Boon, "Developments in the Field of Close Rela-tionships: Creating Foundations for Intervention Strategies," *Personality and So-cial Psychology Bulletin* 16, no. 1 (1990): 23–41, doi:10.1177/0146167290161003.

7. Roy F. Baumeister and Ellen Bratslavsky, "Passion, Intimacy, and Time: Pas-sionate Love as a Function of Change in Intimacy," *Personality and Social Psy-chology Review* 3, no. 1 (1999): 49–67, doi:10.1207/s15327957pspr0301_3.

8. Berscheid and Ammazzalorso, "Emotional Experience in Close Relationships."

9. Brian A. Anderson, "The Attention Habit: How Reward Learning Shapes Attentional Selection," *Annals of the New York Academy of Sciences* 1369, no. 1 (2016): 24–39, doi:10.1111/nyas.12957.

10. Brian A. Anderson, Patryk A. Laurent, and Steven Yantis, "Value-Driven Attentional Capture," *Proceedings of the National Academy of Sciences* 108, no. 25 (2011): 10367–71, doi:10.1073/pnas.1104047108.

11. Brian A. Anderson, "Value-Driven Attentional Priority Is Context Specific," *Psychonomic Bulletin and Review* 22, no. 3 (2015): 750–56, doi:10.3758/s13423-014-0724-0.

12. Interview with Dr. Tania Lisboa, professional cellist and Research Fellow at London's Royal College of Music, November 2, 2017.

13. Lorraine Carli, "NFPA Encourages Testing Smoke Alarms as Daylight Saving Time Begins," National Fire Protection Association, March 6, 2014, https://www.nfpa.org/News-and-Research/News-and-media/Press-Room/News-releases/2014/NFPA-encourages-testing-smoke-alarms-as-Daylight-Saving-Time-begins.

14. Steve Sternberg, "How Many Americans Floss Their Teeth?" *U.S. News and World Report*, May 2, 2016, https://www.usnews.com/news/articles/2016-05-02/how-many-americans-floss-their-teeth.

15. Gaby Judah, Benjamin Gardner, and Robert Aunger, "Forming a Flossing Habit: An Exploratory Study of the Psychological Determinants of Habit Formation," *British Journal of Health Psychology* 18, no. 2 (2013): 338–53, doi:10.1111/j.2044-8287.2012.02086.x.

16. Jennifer S. Labrecque et al., "Habit Slips: When Consumers Unintentionally Resist New Products," *Journal of the Academy of Marketing Science* 45, no. 1 (2017): 119–33, doi:10.1007/s11747-016-0482-9.

17. Labrecque et al.

18. Psychologists reading this might wonder how stacking differs from *implementation intentions*, or "if-then" plans. Implementation intentions tie an intention to a future event, without regard for whether that event is a habit or not. Labrecque et al. (2017) found that such standard implementation intentions did not increase students' use of the laundry product across the four weeks of the study.

19. Labrecque et al., "Habit Slips." This strategy was called *response substitution* in early behavior therapy.

20. Margot Sanger-Katz, "The Decline of 'Big Soda,'" *The New York Times*, October 2, 2015, https://www.nytimes.com/2015/10/04/upshot/soda-industry-struggles-as-consumer-tastes-change.html.

10. Total Control

1. Emma Runnemark, Jonas Hedman, and Xiao Xiao, "Do Consumers Pay More Using Debit Cards Than Cash?" *Electronic Commerce Research and Applications* 14, no. 5 (2015): 285–91, doi:10.1016/j.elerap.2015.03.002.

2. Jonathan Cantor et al., "Five Years Later: Awareness of New York City's Calorie Labels Declined, with No Changes in Calories Purchased," *Health Affairs*

34, no. 11 (2015): 1893–1900, doi:10.1377/hlthaff.2015.0623; Kamila M. Kiszko et al., "The Influence of Calorie Labeling on Food Orders and Consumption: A Review of the Literature," *Journal of Community Health* 39, no. 6 (2014): 1248–69, doi:10.1007/s10900-014-9876-0; Susan E. Sinclair, Marcia Cooper, and Elizabeth D. Mansfield, "The Influence of Menu Labeling on Calories Selected or Consumed: A Systematic Review and Meta-Analysis," *Journal of the Academy of Nutrition and Dietetics* 114, no. 9 (2014): 1375–88, doi:10.1016/j.jand.2014.05.014; although see Natalina Zlatevska, Nico Neumann, and Chris Dubelaar, "Mandatory Calorie Disclosure: A Comprehensive Analysis of Its Effect on Consumers and Retailers," *Journal of Retailing* 94, no. 1 (2018): 89–101, doi:10.1016/j.jretai.2017.09.007.

3. To Dieu-Hang et al., "Household Adoption of Energy and Water-Efficient Appliances: An Analysis of Attitudes, Labelling and Complementary Green Behaviours in Selected OECD Countries," *Journal of Environmental Management* 197 (2017): 140–50, doi:10.1016/j.jenvman.2017.03.070.

4. Allison Aubrey, "More Salt in School Lunch, Less Nutrition Info on Menus: Trump Rolls Back Food Rules," NPR, May 2, 2017, https://www.npr.org/sections/thesalt/2017/05/02/526448646/trump-administration-rolls-back-obama-era-rules-on-calorie-counts-school-lunch.

5. George Loewenstein, Cass R. Sunstein, and Russell Golman, "Disclosure: Psychology Changes Everything," *Annual Review of Economics* 6 (2014): 391–419, doi:10.1146/annurev-economics-080213-041341.

6. Angela L. Duckworth et al., "A Stitch in Time: Strategic Self-Control in High School and College Students," *Journal of Educational Psychology* 108, no. 3 (2016): 329–41, doi:10.1037/edu0000062.

7. Angela L. Duckworth, Tamar Szabó Gendler, and James J. Gross, "Situational Strategies for Self-Control," *Perspectives on Psychological Science* 11, no. 1 (2016): 35–55, doi:10.1177/1745691615623247.

8. Duckworth et al., "A Stitch in Time."

9. Michael R. Ent, Roy F. Baumeister, and Dianne M. Tice, "Trait Self-Control and the Avoidance of Temptation," *Personality and Individual Differences* 74 (2015): 12–15, doi:10.1016/j.paid.2014.09.031.

10. Ent, Baumeister, and Tice.

11. Ent, Baumeister, and Tice.

12. Michelle R. vanDellen et al., "In Good Company: Managing Interpersonal Resources That Support Self-Regulation," *Personality and Social Psychology Bulletin* 41, no. 6 (2015): 869–82, doi:10.1177/0146167215580778.

13. Brian Wansink and Collin R. Payne, "Eating Behavior and Obesity at Chinese Buffets," *Obesity* 16, no. 8 (2008): 1957–60, doi:10.1038/oby.2008.286. Note that these data are from the Corrigendum and verified by http://www.timvanderzee.com/the-wansink-dossier-an-overview.

14. Jen Labrecque, Kristen Lee, and Wendy Wood, "Overthinking Habit" (manuscript under revision, University of Southern California, 2018).

15. Eric A. Thrailkill et al., "Stimulus Control of Actions and Habits: A Role for Reinforcer Predictability and Attention in the Development of Habitual Behavior," *Journal of Experimental Psychology: Animal Learning and Cognition* 44, no. 4 (2018): 370–84, doi:10.1037/xan0000188.

16. Claire M. Gillan et al., "Model-Based Learning Protects against Forming Habits," *Cognitive, Affective, and Behavioral Neuroscience* 15, no. 3 (2015): 523–36, doi:10.3758/s13415-015-0347-6.

11. Jump Through Windows

1. Shaun Larcom, Ferdinand Rauch, and Tim Willems, "The Benefits of Forced Experimentation: Striking Evidence from the London Underground Network," *The Quarterly Journal of Economics* 132, no. 4 (2017): 2019–55, doi:10.1093/qje/qjx020.

2. Bas Verplanken et al., "Context Change and Travel Mode Choice: Combining the Habit Discontinuity and Self-Activation Hypotheses," *Journal of Environmental Psychology* 28, no. 2 (2008): 121–27, doi:10.1016/j.jenvp.2007.10.005.

3. Félix Ravaisson, *Of Habit*, trans. Clare Carlisle and Mark Sinclair (London: Continuum, 2008; orig. pub. 1838).

4. Roy F. Baumeister and Ellen Bratslavsky, "Passion, Intimacy, and Time: Passionate Love as a Function of Change in Intimacy," *Personality and Social Psychology Review* 3, no. 1 (1999): 49–67, doi:10.1207/s15327957pspr0301_3.

5. Baumeister and Bratslavsky.

6. Verplanken et al., "Context Change and Travel Mode Choice."

7. Sam K. Hui et al., "The Effect of In-Store Travel Distance on Unplanned Spending: Applications to Mobile Promotion Strategies," *Journal of Marketing* 77, no. 2 (2013): 1–16, doi:10.1509/jm.11.0436.

8. Hui et al.

9. Tom Ryan, "Older Shoppers Irritated by Supermarket Layout Changes," RetailWire, March 12, 2012, http://www.retailwire.com/discussion/older-shoppers -irritated-by-supermarket-layout-changes/.

10. Scott Young and Vincenzo Ciummo, "Managing Risk in a Package Redesign: What Can We Learn from Tropicana?" *Brand Packaging* (2009): 18–21, https://www.highbeam.com/doc/1G1-208131373.html.

11. David L. Alexander, John G. Lynch Jr., and Qing Wang, "As Time Goes By: Do Cold Feet Follow Warm Intentions for Really New versus Incrementally New Products?," *Journal of Marketing Research* 45, no. 3 (2008): 307–19, https://www.jstor.org/stable/30162533.

12. Matthew Lynley, "Bird Has Officially Raised a Whopping $300M as the Scooter Wars Heat Up," *TechCrunch*, June 28, 2018, https://techcrunch.com/2018/06 /28/bird-has-officially-raised-a-whopping-300m-as-the-scooter-wars-heat-up.

13. Alexander, Lynch, and Wang, "As Time Goes By," 307–19.

14. Thad Dunning et al., "Is Paying Taxes Habit Forming? Experimental Evidence from Uruguay," presentation, University of California, Berkeley, 2017, http://www.thaddunning.com/wp-content/uploads/2017/09/Dunning-et-al_Habit _2017.pdf.

15. Dunning et al., 34.

16. Thomas Fujiwara, Kyle Meng, and Tom Vogl, "Habit Formation in Voting: Evidence from Rainy Elections," *American Economic Journal: Applied Economics* 8, no. 4 (2016): 160–88, doi:10.1257/app.20140533.

17. Wendy Wood, Leona Tam, and Melissa Guerrero Witt, "Changing Circum-

stances, Disrupting Habits," *Journal of Personality and Social Psychology* 88, no. 6 (2005): 918–33, doi:10.1037/0022-3514.88.6.918.

18. Jewel Jordan, "Americans Moving at Historically Low Rates," *Census Bureau Reports*, United States Census Bureau, November 16, 2016, https://www .census.gov/newsroom/press-releases/2016/cb16-189.html.

19. Mona Chalabi, "How Many Times Does the Average Person Move?" *FiveThirty-Eight*, January 29, 2015, https://fivethirtyeight.com/features/how-many-times -the-average-person-moves/.

20. United States Department of Labor, "Employee Tenure Summary," Bureau of Labor Statistics, September 22, 2016, https://www.bls.gov/news.release/tenure .nr0.htm.

21. Todd F. Heatherton and Patricia A. Nichols, "Personal Accounts of Successful Versus Failed Attempts at Life Change," *Personality and Social Psychology Bulletin* 20, no. 6 (1994): 664–75, doi:10.1177/0146167294206005.

22. Bryan L. Rogers et al., "Turning Up by Turning Over: The Change of Scenery Effect in Major League Baseball," *Journal of Business and Psychology* 32, no. 5 (2017): 547–60, doi:10.1007/s10869-016-9468-3.

23. These were two *sabermetrics* (in baseball analytics, composite statistical measures of individual player performance): *on-base plus slugging* (OPS), which reflects a player's ability to reach base and hit for power; *weighted runs created plus* (wRC+), which reflects a player's overall offensive contribution relative to other players. See https://www.fangraphs.com.

12. The Special Resilience of Habit

1. "2015 Stress in America," American Psychological Association, accessed March 13, 2018, http://www.apa.org/news/press/releases/stress/2015/snapshot.aspx.

2. Grant S. Shields, Matthew A. Sazma, and Andrew P. Yonelinas, "The Effects of Acute Stress on Core Executive Functions: A Meta-Analysis and Comparison with Cortisol," *Neuroscience and Biobehavioral Reviews* 68 (2016): 651–68, doi:10.1016/j.neubiorev.2016.06.038.

3. David T. Neal, Wendy Wood, and Aimee Drolet, "How Do People Adhere to Goals When Willpower Is Low? The Profits (and Pitfalls) of Strong Habits," *Journal of Personality and Social Psychology* 104, no. 6 (2013): 959–75, doi: 10.1037/a0032626.

4. Lars Schwabe and Oliver T. Wolf, "Stress Increases Behavioral Resistance to Extinction," *Psychoneuroendocrinology* 36, no. 9 (2011): 1287–93, doi:10.1016/j .psyneuen.2011.02.002.

5. Mike Mannor et al., "How Anxiety Affects CEO Decision Making," Harvard Business Review, July 19, 2016, https://hbr.org/2016/07/how-anxiety-affects-ceo -decision-making.

6. James G. March, "Exploration and Exploitation in Organizational Learning," *Organization Science* 2, no. 1 (1991): 71–87, https://www.jstor.org/stable/2634940.

7. Lars Schwabe and Oliver T. Wolf, "Stress and Multiple Memory Systems: From 'Thinking' to 'Doing,'" *Trends in Cognitive Sciences* 17, no. 2 (2013): 60–68, doi:10.1016/j.tics.2012.12.001.

8. Schwabe and Wolf.

9. Neal, Wood, and Drolet, "How Do People Adhere to Goals When Willpower Is Low?"

10. Neal, Wood, and Drolet.

11. Donald E. Broadbent et al., "The Cognitive Failures Questionnaire (CFQ) and Its Correlates," *British Journal of Clinical Psychology* 21, no. 1 (1982): 1–16, doi:10.1111/j.2044-8260.1982.tb01421.x.

12. María K. Jónsdóttir et al., "A Diary Study of Action Slips in Healthy Individuals," *Clinical Neuropsychologist* 21, no. 6 (2007): 875–83, doi:10.1080/13854040701220044.

13. Rachel J. Katz-Sidlow et al., "Smartphone Use During Inpatient Attending Rounds: Prevalence, Patterns and Potential for Distraction," *Journal of Hospital Medicine* 7, no. 8 (2012): 595–99, doi:10.1002/jhm.1950.

14. Trevor Smith, Edward Darling, and Bruce Searles, "2010 Survey on Cell Phone Use While Performing Cardiopulmonary Bypass," *Perfusion* 26, no. 5 (2011): 375–80, doi:10.1177/0267659111409969.

15. Jack L. Nasar and Derek Troyer, "Pedestrian Injuries Due to Mobile Phone Use in Public Places," *Accident Analysis and Prevention* 57 (2013): 91–95, doi:10.1016/j.aap.2013.03.021.

16. James Reason and Deborah Lucas, "Absent-Mindedness in Shops: Its Incidence, Correlates and Consequences," *British Journal of Clinical Psychology* 23, no. 2 (1984): 121–31, doi:10.1111/j.2044-8260.1984.tb00635.x.

17. Reason and Lucas.

18. Arun Vishwanath, "Examining the Distinct Antecedents of E-Mail Habits and Its Influence on the Outcomes of a Phishing Attack," *Journal of Computer-Mediated Communication* 20, no. 5 (2015): 570–84, doi:10.1111/jcc4.12126.

19. Arun Vishwanath, "Habitual Facebook Use and Its Impact on Getting Deceived on Social Media," *Journal of Computer-Mediated Communication* 20, no. 1 (2015): 83–98, doi:10.1111/jcc4.12100.

20. Mathew A. Harris and Thomas Wolbers, "How Age-Related Strategy Switching Deficits Affect Wayfinding in Complex Environments," *Neurobiology of Aging* 35, no. 5 (2014): 1095–1102, doi:10.1016/j.neurobiolaging.2013.10.086.

13. Contexts of Addiction

1. "Drugs, Brains, and Behavior: The Science of Addiction," National Institute on Drug Abuse, last modified July 2018, https://www.drugabuse.gov/publications/drugs-brains-behavior-science-addiction/drug-abuse-addiction.

2. Aldo Badiani et al., "Opiate versus Psychostimulant Addiction: The Differences Do Matter," *Nature Reviews Neuroscience* 12, no. 11 (2011): 685–700, doi:10.1038/nrn3104; Aldo Badiani et al., "Addiction Research and Theory: A Commentary on the Surgeon General's Report on Alcohol, Drugs, and Health," *Addiction Biology* 23, no. 1 (2018): 3–5, doi:10.1111/adb.12497.

3. David J. Nutt et al., "The Dopamine Theory of Addiction: 40 Years of Highs and Lows," *Nature Reviews Neuroscience* 16, no. 5 (2015): 305–12, doi:10.1038/nrn3939.

4. Kent C. Berridge and Terry E. Robinson, "Liking, Wanting, and the Incentive-

Sensitization Theory of Addiction," *American Psychologist* 71, no. 8 (2016): 670–79, doi:10.1037/amp0000059.

5. Barry J. Everitt and Trevor W. Robbins, "Drug Addiction: Updating Actions to Habits to Compulsions Ten Years On," *Annual Review of Psychology* 67, no. 1 (2016): 23–50, doi:10.1146/annurev-psych-122414-033457.

6. Rebecca Ahrnsbrak et al., *Key Substance Use and Mental Health Indicators in the United States: Results from the 2016 National Survey on Drug Use and Health*, HHS Publication No. SMA 17-5044, NSDUH Series H-52 (Rockville, MD: Center for Behavioral Health Statistics and Quality, Substance Abuse and Mental Health Services Administration, 2017); "Alcohol Use: Data and Statistics," World Health Organization, accessed February 16, 2019, http://www.euro.who.int/en/health-topics/disease-prevention/alcohol-use/data-and-statistics.

7. Eunice Park-Lee et al., *Receipt of Services for Substance Use and Mental Health Issues Among Adults: Results from the 2016 National Survey on Drug Use and Health* (Rockville, MD: SAMHSA: NSDUH Data Review, September 2017).

8. "Drugs, Brains, and Behavior," National Institute on Drug Abuse.

9. Paul Crits-Christoph et al., "Psychosocial Treatments for Cocaine Dependence: National Institute on Drug Abuse Collaborative Cocaine Treatment Study," *Archives of General Psychiatry* 56, no. 6 (1999): 493–502.

10. James R. McKay, "Making the Hard Work of Recovery More Attractive for Those with Substance Use Disorders," *Addiction* 112, no. 5 (2017): 751–57, doi:10.1111/add.13502.

11. Alvin M. Shuster, "G.I. Heroin Addiction Epidemic in Vietnam," *The New York Times*, May 16, 1971, http://www.nytimes.com/1971/05/16/archives/gi-heroin-addiction-epidemic-in-vietnam-gi-heroin-addiction-is.html.

12. Jeremy Kuzmarov, *The Myth of the Addicted Army: Vietnam and the Modern War on Drugs* (Amherst, MA: University of Massachusetts Press, 2009).

13. Lee N. Robins et al., "Vietnam Veterans Three Years After Vietnam: How Our Study Changed Our View of Heroin," *American Journal on Addiction* 19, no. 3 (2010): 203–11, doi:10.1111/j.1521-0391.2010.00046.x; Lee N. Robins, "Vietnam Veterans' Rapid Recovery from Heroin Addiction: A Fluke or Normal Expectation?" *Addiction* 88, no. 8 (1993): 1041–54, doi:10.1111/j.1360-0443.1993.tb02123.x.

14. Interview with Vietnam War veteran, December, 9, 2017. The name of the interviewee is withheld for confidentiality.

15. Lee N. Robins, Darlene H. Davis, and Donald W. Goodwin, "Drug Use by US Army Enlisted Men in Vietnam: A Follow-up on Their Return Home," *American Journal of Epidemiology* 99, no. 4 (1974): 235–49, doi:10.1093/oxfordjournals.aje.a121608.

16. Robins, "Vietnam Veterans' Rapid Recovery from Heroin Addiction."

17. Robins et al., "Vietnam Veterans Three Years After Vietnam: How Our Study Changed Our View of Heroin."

18. Robins, "Vietnam Veterans' Rapid Recovery from Heroin Addiction."

19. Alix Spiegel, "What Vietnam Taught Us About Breaking Bad Habits," NPR, January 2, 2012, http://www.npr.org/sections/health-shots/2012/01/02/144431794/what-vietnam-taught-us-about-breaking-bad-habits.

20. Robins et al., "Vietnam Veterans Three Years After Vietnam."

21. Robins, "Vietnam Veterans' Rapid Recovery from Heroin Addiction," 1046.

22. Robins, 1046.

23. Robins, 1031.

24. Patricia F. Hadaway et al., "The Effect of Housing and Gender on Preference for Morphine-Sucrose Solutions in Rats," *Psychopharmacology* 66, no. 1 (1979): 87–91, doi:10.1007/bf00431995; Bruce K. Alexander et al., "Effect of Early and Later Colony Housing on Oral Ingestion of Morphine in Rats," *Pharmacology Biochemistry and Behavior* 15, no. 4 (1981): 571–76, doi:10.1016/0091-3057 (81)90211-2; Rebecca S. Hofford et al., "Effects of Environmental Enrichment on Self-Administration of the Short-Acting Opioid Remifentanil in Male Rats," *Psychopharmacology* 234, no. 23–24 (2017): 3499–506, doi:10.1007/s00213 -017-4734-2.

25. Kenneth J. Thiel et al., "Anti-Craving Effects of Environmental Enrichment," *International Journal of Neuropsychopharmacology* 12, no. 9 (2009): 1151–56, doi:10.1017/S1461145709990472; see also Seven E. Tomek and M. Foster Olive, "Social Influences in Animal Models of Opiate Addiction," *International Review of Neurobiology* 140 (2018): 81–107, doi:10.1016/bs.irn.2018.07.004; Ewa Galaj, Monica Manuszak, and Robert Ranaldi, "Environmental Enrichment as a Potential Intervention for Heroin Seeking," *Drug and Alcohol Dependence* 163 (2016): 195–201, doi:10.1016/j.drugalcdep.2016.04.016.

26. Bruce K. Alexander and Patricia F. Hadaway, "Opioid Addiction: The Case for an Adaptive Orientation," *Psychological Bulletin* 92, no. 2 (1982): 367–81, doi:10.1037/0033-2909.92.2.367.

27. "Drugs, Brains, and Behavior," National Institute on Drug Abuse.

28. Anke Snoek, Neil Levy, and Jeanette Kennett, "Strong-Willed but Not Successful: The Importance of Strategies in Recovery from Addiction," *Addictive Behaviors Reports* 4 (2016): 102–7, doi:10.1016/j.abrep.2016.09.002.

29. Snoek, Levy, and Kennett, 107.

30. Jenna Payesko, "FDA Approves Lofexidine Hydrochloride, First Non-Opioid Treatment for Management of Opioid Withdrawal Symptoms in Adults," *Med Magazine*, May 16, 2018, https://www.mdmag.com/medical-news/fda-approves -lofexidine-hydrochloride-first-nonopioid-treatment-for-management-of-opioid -withdrawal-symptoms-in-adults.

31. McKay, "Making the Hard Work of Recovery More Attractive," 752.

32. McKay, 752.

33. George M. Hunt and Nathan H. Azrin, "A Community-Reinforcement Approach to Alcoholism," *Behaviour Research and Therapy* 11, no. 1 (1973): 91–104, doi:10.1016/0005-7967(73)90072-7.

34. Kenneth Silverman, Anthony DeFulio, and Sigurdur O. Sigurdsson, "Maintenance of Reinforcement to Address the Chronic Nature of Drug Addiction," *Preventive Medicine* 55 (2012): S46–S53, doi:10.1016/j.ypmed.2012.03.013.

35. Mark D. Litt et al., "Changing Network Support for Drinking: Network Support Project 2-Year Follow-up," *Journal of Consulting and Clinical Psychology* 77, no. 2 (2009): 229–42, doi:10.1037/a0015252.

36. Silverman, DeFulio, and Sigurdsson, "Maintenance of Reinforcement to Address the Chronic Nature of Drug Addiction."

37. Anthony DeFulio and Kenneth Silverman, "Employment-Based Abstinence Reinforcement as a Maintenance Intervention for the Treatment of Cocaine

Dependence: Post-intervention Outcomes," *Addiction* 106, no. 5 (2011): 960–67, doi:10.1111/j.1360-0443.2011.03364.x.

38. Kenneth Silverman, August F. Holtyn, and Reed Morrison, "The Therapeutic Utility of Employment in Treating Drug Addiction: Science to Application," *Translational Issues in Psychological Science* 2, no. 2 (2016): 203–12, doi:10.1037/tps0000061.

39. George F. Koob and Nora D. Volkow, "Neurobiology of Addiction: A Neurocircuitry Analysis," *The Lancet Psychiatry* 3, no. 8 (2016): 760–73, doi:10.1016/S2215-0366(16)00104-8.

40. John Monterosso and Wendy Wood, "Habits of Successful Rehabilitation" (unpublished data, University of Southern California, 2017).

14. Happy with Habit

1. Samantha J. Caton et al., "Repetition Counts: Repeated Exposure Increases Intake of a Novel Vegetable in UK Pre-School Children Compared to Flavour–Flavour and Flavour–Nutrient Learning," *British Journal of Nutrition* 109, no. 11 (2013): 2089–97, doi:10.1017/s0007114512004126.

2. Edward Bradford Titchener, *A Textbook of Psychology*, rev. ed. (New York: Macmillan, 1896, repr. 1928), 408.

3. Robert B. Zajonc, "Attitudinal Effects of Mere Exposure," *Journal of Personality and Social Psychology* 9, no. 2 (1968): 1–27, doi:10.1037/h0025848.

4. Robert F. Bornstein and Catherine Craver-Lemley, "Mere Exposure Effect," in *Cognitive Illusions: Intriguing Phenomena in Thinking, Judgment and Memory*, ed. Rüdiger F. Pohl, 2nd ed. (New York: Routledge, 2017), 256–75.

5. Theodore H. Mita, Marshall Dermer, and Jeffrey Knight, "Reversed Facial Images and the Mere-Exposure Hypothesis," *Journal of Personality and Social Psychology* 35, no. 8 (1977): 597–601, doi:10.1037//0022-3514.35.8.597.

6. Rolf Reber, Norbert Schwarz, and Piotr Winkielman, "Processing Fluency and Aesthetic Pleasure: Is Beauty in the Perceiver's Processing Experience?" *Personality and Social Psychology Review* 8, no. 4 (2004): 364–82, doi:10.1207/s15327957pspr0804_3.

7. Christian Obermeier et al., "Aesthetic Appreciation of Poetry Correlates with Ease of Processing in Event-Related Potentials," *Cognitive, Affective, and Behavioral Neuroscience* 16, no. 2 (2016): 362–73, doi:10.3758/s13415-015-0396-x.

8. Stefan Mayer and Jan R. Landwehr, "Objective Measures of Design Typicality," *Design Studies* 54 (2018): 146–61, doi:10.1016/j.destud.2017.09.004; Stefan Mayer and Jan R. Landwehr, "Objective Measures of Design Typicality That Predict Aesthetic Liking, Fluency, and Car Sales," in *Advances in Consumer Research* 44 (Duluth, MN: Association for Consumer Research, 2016): 556–57.

9. Susanne Schmidt and Martin Eisend, "Advertising Repetition: A Meta-Analysis on Effective Frequency in Advertising," *Journal of Advertising* 44, no. 4 (2015): 415–28, doi:10.1080/00913367.2015.1018460; R. Matthew Montoya et al., "A Re-Examination of the Mere Exposure Effect: The Influence of Repeated Exposure on Recognition, Familiarity, and Liking," *Psychological Bulletin* 143, no. 5 (2017): 459–98, doi:10.1037/bul0000085.

10. Thomas N. Robinson et al., "Effects of Fast Food Branding on Young Children's Taste Preferences," *Archives of Pediatrics and Adolescent Medicine* 161, no. 8 (2007): 792–97, doi:10.1001/archpedi.161.8.792.

11. Dinah Avni-Babad, "Routine and Feelings of Safety, Confidence, and Well-Being," *British Journal of Psychology* 102, no. 2 (2011): 223–44, doi:10.1348/000712610x513617.

12. Avni-Babad.

13. Avni-Babad.

14. Richard Florida, "The Geography of Car Deaths in America," *CityLab*, October 15, 2015, http://www.citylab.com/commute/2015/10/the-geography-of-car-deaths-in-america/410494.

15. Mindy F. Ji and Wendy Wood, "Purchase and Consumption Habits: Not Necessarily What You Intend," *Journal of Consumer Psychology* 17, no. 4 (2007): 261–76, doi:10.1016/S1057-7408(07)70037-2.

16. Clifford Geertz, *The Interpretation of Cultures* (New York: Basic Books, 1973).

17. Allen Ding Tian et al., "Enacting Rituals to Improve Self-Control," *Journal of Personality and Social Psychology* 114, no. 6 (2018): 851–76, doi:10.1037/pspa0000113.

18. Michaéla C. Schippers and Paul A. M. Van Lange, "The Psychological Benefits of Superstitious Rituals in Top Sport: A Study Among Top Sportspersons," *Journal of Applied Social Psychology* 36, no. 10 (2006): 2532–53, doi:10.1111/j.0021-9029.2006.00116.x.

19. Nicholas M. Hobson, Devin Bonk, and Michael Inzlicht, "Rituals Decrease the Neural Response to Performance Failure," *PeerJ* 5 (2017): e3363, doi:10.7717/peerj.3363.

20. Cristine H. Legare and André L. Souza, "Evaluating Ritual Efficacy: Evidence from the Supernatural," *Cognition* 124, no. 1 (2012): 1–15, doi:10.1016/j.cognition.2012.03.004.

21. Michael I. Norton and Francesca Gino, "Rituals Alleviate Grieving for Loved Ones, Lovers, and Lotteries," *Journal of Experimental Psychology: General* 143, no. 1 (2014): 266–72, doi:10.1037/a0031772.

22. Alison Wood Brooks et al., "Don't Stop Believing: Rituals Improve Performance by Decreasing Anxiety," *Organizational Behavior and Human Decision Processes* 137 (2016): 71–85, doi:10.1016/j.obhdp.2016.07.004.

23. Norton and Gino, "Rituals Alleviate Grieving for Loved Ones, Lovers, and Lotteries."

24. Brooks et al., "Don't Stop Believing."

25. Daniel L. Wann et al., "Examining the Superstitions of Sport Fans: Types of Superstitions, Perceptions of Impact, and Relationship with Team Identification," *Athletic Insight* 5, no. 1 (2013): 21–44. Retrieved from http://libproxy.usc.edu/login?url=https://search.proquest.com/docview/1623315047?accountid=14749.

26. Mihaly Csikszentmihalyi, *Flow: The Psychology of Optimal Experience* (New York: Harper Perennial, 1996).

27. Samantha J. Heintzelman and Laura A. King, "Routines and Meaning in Life," *Personality and Social Psychology Bulletin* (published online September 18, 2018): doi:10.1177/0146167218795133.

28. Matthew Hutson, "Everyday Routines Make Life Feel More Meaningful," *Scientific American*, July 1, 2015, https://www.scientificamerican.com/article/everyday-routines-make-life-feel-more-meaningful/.

29. Aditi Shrikant, "11 Senior Citizens on the Best Products of the Past Century," *Vox*, December 11, 2018, https://www.vox.com/the-goods/2018/12/11/18116313/best-products-seniors-elderly-tide-samsung.

15. You Are Not Alone

1. Richard H. Thaler and Cass R. Sunstein, *Nudge: Improving Decisions about Health, Wealth, and Happiness*, updated ed. (New York: Penguin, 2009), 8.

2. Lee Shepherd, Ronan E. O'Carroll, and Eamonn Ferguson, "An International Comparison of Deceased and Living Organ Donation/Transplant Rates in Opt-In and Opt-Out Systems: A Panel Study," *BMC Medicine* 12, no. 1 (2014): 131, doi:10.1186/s12916-014-0131-4.

3. Shlomo Benartzi, "Save More Tomorrow," 2017, http://www.shlomobenartzi.com/save-more-tomorrow.

4. "2014 State Indicator Report on Physical Activity," Centers for Disease Control and Prevention (Atlanta, GA: U.S. Department of Health and Human Services, 2014), https://www.cdc.gov/physicalactivity/downloads/pa_state_indicator_report_2014.pdf.

5. Molly Warren, Stacy Beck, and Jack Rayburn, *The State of Obesity: Better Policies for a Healthier America 2018* (Washington, D.C.: Trust for America's Health, 2018), 1–68.

6. Mariana Arcaya et al., "Urban Sprawl and Body Mass Index among Displaced Hurricane Katrina Survivors," *Preventive Medicine* 65 (2014): 40–46, doi:10.1016/j.ypmed.2014.04.006; see also Jana A. Hirsch et al., "Change in Walking and Body Mass Index Following Residential Relocation: The Multi-Ethnic Study of Atherosclerosis," *American Journal of Public Health* 104, no. 3 (2014): e49–e56, doi:10.2105/ajph.2013.301773.

7. Adam Martin et al., "Impact of Changes in Mode of Travel to Work on Changes in Body Mass Index: Evidence from the British Household Panel Survey," *Journal of Epidemiology and Community Health* 69, no. 8 (2015): 753–61, doi:10.1136/jech-2014-205211.

8. Matthew Hall, "Bird Scooters Flying Around Town," *Santa Monica Daily Press*, September 26, 2017, http://smdp.com/bird-scooters-flying-around-town/162647.

9. National Association of City Transportation Officials, *Equitable Bike Share Means Building Better Places for People to Ride*, July 2016, https://nacto.org/wp-content/uploads/2016/07/NACTO_Equitable_Bikeshare_Means_Bike_Lanes.pdf.

10. NYC DOT, *Cycling in the City: Cycling Trends in NYC*, 2018, http://www.nyc.gov/html/dot/downloads/pdf/cycling-in-the-city.pdf.

11. Allison B. Brenner et al., "Longitudinal Associations of Neighborhood Socioeconomic Characteristics and Alcohol Availability on Drinking: Results from the Multi-Ethnic Study of Atherosclerosis (MESA)," *Social Science and Medicine* 145 (2015): 17–25, doi:10.1016/j.socscimed.2015.09.030; see also Sarah Foster

et al., "Liquor Landscapes: Does Access to Alcohol Outlets Influence Alcohol Consumption in Young Adults?," *Health and Place* 45 (2017): 17–23, doi:10.1016/j .healthplace.2017.02.008.

12. Hunter Schwarz, "Where in the United States You Can't Purchase Alcohol," *The Washington Post*, September 2, 2014, https://www.washingtonpost.com /blogs/govbeat/wp/2014/09/02/where-in-the-united-states-you-cant-purchase -alcohol.

13. Jennifer Ahern et al., "Alcohol Outlets and Binge Drinking in Urban Neighborhoods: The Implications of Nonlinearity for Intervention and Policy," *American Journal of Public Health* 103, no. 4 (2013): e81–e87, doi:10.2105/ajph.2012 .301203.

14. Michael Pollan, "The Way We Live Now: 10-12-03; The (Agri)Cultural Contradictions of Obesity," *The New York Times Magazine*, October 12, 2003, http:// www.nytimes.com/2003/10/12/magazine/the-way-we-live-now-10-12-03-the -agri-cultural-contradictions-of-obesity.html.

15. "Portion Distortion," National Heart, Lung, and Blood Institute, last modified on April 1, 2015, https://www.nhlbi.nih.gov/health/educational/wecan/eat-right /portion-distortion.htm.

16. Gareth J. Hollands et al., "Portion, Package or Tableware Size for Changing Selection and Consumption of Food, Alcohol and Tobacco," *Cochrane Database of Systematic Reviews* 9, no. CD011045 (2015): https://www.ncbi.nlm.nih.gov /pmc/articles/PMC4579823/; Natalina Zlatevska, Chris Dubelaar, and Stephen S. Holden, "Sizing Up the Effect of Portion Size on Consumption: A Meta-Analytic Review," *Journal of Marketing* 78, no. 3 (2014): 140–54, doi:10.1509 /jm.12.0303.

17. Margot Sanger-Katz, "Yes, Soda Taxes Seem to Cut Soda Drinking," *The New York Times*, October 13, 2015, https://www.nytimes.com/2015/10/13/upshot/yes -soda-taxes-seem-to-cut-soda-drinking.html.

18. Lynn D. Silver et al., "Changes in Prices, Sales, Consumer Spending, and Beverage Consumption One Year after a Tax on Sugar-Sweetened Beverages in Berkeley, California, US: A Before-and-After Study," *PLoS Medicine* 14, no. 4 (2017): e1002283, doi:10.1371/journal.pmed.1002283.

19. M. Arantxa Colchero et al., "In Mexico, Evidence of Sustained Consumer Response Two Years After Implementing a Sugar-Sweetened Beverage Tax," *Health Affairs* 36, no. 3 (2017): 564–71, doi:10.1377/hlthaff.2016.1231.

20. Lindsey Smith Taillie et al., "Do High vs. Low Purchasers Respond Differently to a Nonessential Energy-Dense Food Tax? Two-Year Evaluation of Mexico's 8% Nonessential Food Tax," *Preventive Medicine* 105 (2017): S37–S42, doi:10.1016 /j.ypmed.2017.07.009.

21. Drew DeSilver, "Perceptions and Realities of Recycling Vary Widely from Place to Place," Pew Research Center, October 7, 2016, http://www.pewresearch.org /fact-tank/2016/10/07/perceptions-and-realities-of-recycling-vary-widely-from -place-to-place.

22. Adam Cooper, "Electric Company Smart Meter Deployments: Foundation for a Smart Grid," Institute for Electric Innovation, December 2017, http://www .edisonfoundation.net/iei/publications/Documents/IEI_Smart%20Meter%20 Report%202017_FINAL.pdf.

23. Chris Mooney, "Why 50 Million Smart Meters Still Haven't Fixed America's Energy Habits," *The Washington Post*, January 29, 2015, https://www.washingtonpost.com/news/energy-environment/wp/2015/01/29/americans-are-this-close-to-finally-understanding-their-electricity-bills.

24. Katrina Jessoe and David Rapson, "Knowledge Is (Less) Power: Experimental Evidence from Residential Energy Use," *American Economic Review* 104, no.4 (2014): 1417–38, doi:10.1257/aer.104.4.1417.

How to Stop Looking at Your Phone So Often

1. Frank Newport, "Email Outside of Working Hours Not a Burden to U.S. Workers," Gallup, May 10, 2017, https://news.gallup.com/poll/210074/email-outside-working-hours-not-burden-workers.aspx.

2. Jan Dettmers, "How Extended Work Availability Affects Well-Being: The Mediating Roles of Psychological Detachment and Work-Family Conflict," *Work and Stress* 31, no. 1 (2017): 24–41, doi:10.1080/02678373.2017.1298164; Jim Harter, "Should Employers Ban Email After Work Hours?" Gallup, September 9, 2014, https://www.gallup.com/workplace/236519/employers-ban-email-work-hours.aspx.

3. Jan Dettmers et al., "Extended Work Availability and Its Relation with Start-of-Day Mood and Cortisol," *Journal of Occupational Health Psychology* 21, no. 1 (2016): 105–18, doi:10.1037/a0039602.

4. Cary Stothart, Ainsley Mitchum, and Courtney Yehnert, "The Attentional Cost of Receiving a Cell Phone Notification," *Journal of Experimental Psychology: Human Perception and Performance* 41, no. 4 (2015): 893–97, doi:10.1037/xhp0000100.

5. James A. Roberts and Meredith E. David, "My Life Has Become a Major Distraction from My Cell Phone: Partner Phubbing and Relationship Satisfaction among Romantic Partners," *Computers in Human Behavior* 54 (2016): 134–41, doi:10.1016/j.chb.2015.07.058; Brandon T. McDaniel and Sarah M. Coyne, "'Technoference': The Interference of Technology in Couple Relationships and Implications for Women's Personal and Relational Well-Being," *Psychology of Popular Media Culture* 5, no. 1 (2016): 85–98, doi:10.1037/ppm0000065.

6. Daniel Halpern and James E. Katz, "Texting's Consequences for Romantic Relationships: A Cross-Lagged Analysis Highlights Its Risks," *Computers in Human Behavior* 71 (2017): 386–94. doi:10.1016/j.chb.2017.01.051.

Bibliography

Aarts, Henk, Bas Verplanken, and Ad van Knippenberg. "Habit and Information Use in Travel Mode Choices." *Acta Psychologica* 96, nos. 1–2 (1997): 1–14. https://doi.org/10.1016/s0001-6918(97)00008-5.

Adams, Christopher D. "Variations in the Sensitivity of Instrumental Responding to Reinforcer Devaluation." *Quarterly Journal of Experimental Psychology Section 34B*, no. 2b (1982): 77–98. https://doi.org/10.1080/14640748208400878.

Adams, Christopher D., and Anthony Dickinson. "Instrumental Responding Following Reinforcer Devaluation." *Quarterly Journal of Experimental Psychology Section 33B*, no. 2 (1981): 109–21. https://doi.org/10.1080/14640748108400816.

Ahern, Jennifer, Claire Margerison-Zilko, Alan Hubbard, and Sandro Galea. "Alcohol Outlets and Binge Drinking in Urban Neighborhoods: The Implications of Nonlinearity for Intervention and Policy." *American Journal of Public Health* 103, no. 4 (2013): e81–e87. https://doi.org/10.2105/ajph.2012.301203.

Ahrnsbrak, Rebecca, Jonaki Bose, Sarra L. Hedden, Rachel N. Lipari, and Eunice Park-Lee. *Key Substance Use and Mental Health Indicators in the United States: Results from the 2016 National Survey on Drug Use and Health.* Rockville, MD: Center for Behavioral Health Statistics and Quality, Substance Abuse and Mental Health Services Administration, 2017.

Ajzen, Icek. "Residual Effects of Past on Later Behavior: Habituation and Reasoned Action Perspectives." *Personality and Social Psychology Review* 6, no. 2 (2002): 107–22. https://doi.org/10.1207/S15327957PSPR0602_02.

"Alcohol Use: Data and Statistics." World Health Organization. Accessed February 16, 2019. http://www.euro.who.int/en/health-topics/disease-prevention/alcohol-use/data-and-statistics.

Aldrich, John H., Jacob M. Montgomery, and Wendy Wood. "Turnout as a Habit." *Political Behavior* 33, no. 4 (2011): 535–63. https://doi.org/10.1007/s11109-010-9148-3.

Alexander, Bruce K., Barry L. Beyerstein, Patricia F. Hadaway, and Robert B. Coambs. "Effect of Early and Later Colony Housing on Oral Ingestion of Morphine in Rats." *Pharmacology Biochemistry and Behavior* 15, no. 4 (1981): 571–76. https://doi.org/10.1016/0091-3057(81)90211-2.

Alexander, Bruce K., and Patricia F. Hadaway. "Opioid Addiction: The Case for an Adaptive Orientation." *Psychological Bulletin* 92, no. 2 (1982): 367–81. https://doi.org/10.1037/0033-2909.92.2.367.

Alexander, David L., John G. Lynch, and Qing Wang. "As Time Goes By: Do Cold Feet Follow Warm Intentions for Really New Versus Incrementally New Products?" *Journal of Marketing Research* 45, no. 3 (2008): 307–19. https://www.jstor.org/stable/30162533.

American Psychological Association. "2015 Stress in America." Accessed March 13, 2018. http://www.apa.org/news/press/releases/stress/2015/snapshot.aspx.

Amodio, David M. "Social Cognition 2.0: An Interactive Memory Systems Account." *Trends in Cognitive Sciences* 23, no. 1 (2018): 21–33. https://doi.org/10.1016/j.tics.2018.10.002.

Anderson, Brian A. "The Attention Habit: How Reward Learning Shapes Attentional Selection." *Annals of the New York Academy of Sciences* 1369, no. 1 (2016): 24–39. https://doi.org/10.1111/nyas.12957.

———. "Value-Driven Attentional Priority Is Context Specific." *Psychonomic Bulletin and Review* 22, no. 3 (2015): 750–56. https://doi.org/10.3758/s13423-014-0724-0.

Anderson, Brian A., Patryk A. Laurent, and Steven Yantis. "Value-Driven Attentional Capture." *Proceedings of the National Academy of Sciences* 108, no. 25 (2011): 10367–71. https://doi.org/10.1073/pnas.1104047108.

Anselme, Patrick. "Dopamine, Motivation, and the Evolutionary Significance of Gambling-Like Behaviour." *Behavioural Brain Research* 256 (2013): 1–4. https://doi.org/10.1016/j.bbr.2013.07.039.

Arcaya, Mariana, Peter James, Jean E. Rhodes, Mary C. Waters, and S. V. Subramanian. "Urban Sprawl and Body Mass Index Among Displaced Hurricane Katrina Survivors." *Preventive Medicine* 65 (2014): 40–46. https://doi.org/10.1016/j.ypmed.2014.04.006.

Ariely, Dan, and Klaus Wertenbroch. "Procrastination, Deadlines, and Performance: Self-Control by Precommitment." *Psychological Science* 13, no. 3 (2002): 219–24. https://doi.org/10.1111/1467-9280.00441.

Armitage, Christopher J. "Can the Theory of Planned Behavior Predict the Maintenance of Physical Activity?" *Health Psychology* 24, no. 3 (2005): 235–45. https://doi.org/10.1037/0278-6133.24.3.235.

Aubrey, Allison. "More Salt in School Lunch, Less Nutrition Info on Menus: Trump Rolls Back Food Rules." NPR. May 2, 2017. https://www.npr.org/sections/thesalt/2017/05/02/526448646/trump-administration-rolls-back-obama-era-rules-on-calorie-counts-school-lunch.

Avni-Babad, Dinah. "Routine and Feelings of Safety, Confidence, and Well-Being." *British Journal of Psychology* 102, no. 2 (2011): 223–44. https://doi.org/10.1348/000712610x513617.

Bachman, Rachel. "How Close Do You Need to Be to Your Gym?" *The Wall Street*

Journal. March 21, 2017. https://www.wsj.com/articles/how-close-do-you-need
-to-be-to-your-gym-1490111186.

Badiani, Aldo, David Belin, David Epstein, Donna Calu, and Yavin Shaham. "Opiate
Versus Psychostimulant Addiction: The Differences Do Matter." *Nature Reviews Neuroscience* 12, no. 11 (2011): 685–700. https://doi.org/10.1038/nrn3104.

Badiani, Aldo, Kent C. Berridge, Markus Heilig, David J. Nutt, and Terry E. Robinson. "Addiction Research and Theory: A Commentary on the Surgeon
General's Report on Alcohol, Drugs, and Health." *Addiction Biology* 23, no. 1
(2018): 3–5. https://doi.org/10.1111/adb.12497.

Baer, Drake. "The Scientific Reason Why Barack Obama and Mark Zuckerberg
Wear the Same Outfit Every Day." *Business Insider.* April 28, 2015. http://www
.businessinsider.com/barack-obama-mark-zuckerberg-wear-the-same-outfit
-2015-4.

Balleine, Bernard W., and John P. O'Doherty. "Human and Rodent Homologies in
Action Control: Corticostriatal Determinants of Goal-Directed and Habitual
Action." *Neuropsychopharmacology* 35, no. 1 (2010): 48–69. https://doi.org/10
.1038/npp.2009.131.

Bargh, John A. *Before You Know It: The Unconscious Reasons We Do What We Do.*
New York: Touchstone, 2017.

Baumeister, Roy F., and Ellen Bratslavsky. "Passion, Intimacy, and Time: Passionate
Love as a Function of Change in Intimacy." *Personality and Social Psychology
Review* 3, no. 1 (1999): 49–67. https://doi.org/10.1207/s15327957pspr0301_3.

Benartzi, Shlomo. "Save More Tomorrow." 2017. http://www.shlomobenartzi.com
/save-more-tomorrow.

Berridge, Kent C., and Terry E. Robinson. "Liking, Wanting, and the Incentive-
Sensitization Theory of Addiction." *American Psychologist* 71, no. 8 (2016): 670–79.
https://doi.org/10.1037/amp0000059.

Berscheid, Ellen, and Hilary Ammazzalorso. "Emotional Experience in Close Relationships." In *Blackwell Handbook of Social Psychology: Interpersonal Processes*,
edited by Garth Fletcher and Margaret Clark. Malden, MA: Blackwell Publishers, 2001.

Berscheid, Ellen, and Pamela Regan. *The Psychology of Interpersonal Relationships.*
New York: Pearson, 2005. Reprint, New York: Routledge, 2016.

Bodor, J. Nicholas, Donald Rose, Thomas A. Farley, Christopher Swalm, and Susanne K. Scott. "Neighbourhood Fruit and Vegetable Availability and Consumption: The Role of Small Food Stores in an Urban Environment." *Public
Health Nutrition* 11, no. 404 (2008): 413–20. https://doi.org/10.1017/s136898
0007000493.

Bornstein, Robert F., and Catherine Craver-Lemley. "Mere Exposure Effect." In
Cognitive Illusions: Intriguing Phenomena in Thinking, Judgment and Memory,
edited by Rüdiger F. Pohl, 256–75. New York: Routledge, 2017.

Brenner, Allison B., Luisa N. Borrell, Tonatiuh Barrientos-Gutierrez, and Ana V.
Diez Roux. "Longitudinal Associations of Neighborhood Socioeconomic Characteristics and Alcohol Availability on Drinking: Results from the Multi-
Ethnic Study of Atherosclerosis (MESA)." *Social Science and Medicine* 145
(2015): 17–25. https://doi.org/10.1016/j.socscimed.2015.09.030.

Broadbent, Donald E., P. Fitzgerald Cooper, Paul FitzGerald, and Katherine R. Parkes. "The Cognitive Failures Questionnaire (CFQ) and Its Correlates." *British Journal of Clinical Psychology* 21, no. 1 (1982): 1–16. https://doi.org/10.1111/j.2044-8260.1982.tb01421.x.

Broers, Valérie J. V., Céline De Breucker, Stephan van den Broucke, and Olivier Luminet. "A Systematic Review and Meta-Analysis of the Effectiveness of Nudging to Increase Fruit and Vegetable Choice." *European Journal of Public Health* 27, no. 5 (2017): 912–20. https://doi.org/10.1093/eurpub/ckx085.

Brooks, Alison Wood, Juliana Schroeder, Jane L. Risen, Francesca Gino, Adam D. Galinsky, Michael I. Norton, and Maurice E. Schweitzer. "Don't Stop Believing: Rituals Improve Performance by Decreasing Anxiety." *Organizational Behavior and Human Decision Processes* 137 (2016): 71–85. https://doi.org/10.1016/j.obhdp.2016.07.004.

Brumage, Jody. "The Public Health Cigarette Smoking Act of 1970." Robert C. Byrd Center. July 25, 2017. https://www.byrdcenter.org/byrd-center-blog/the-public-health-cigarette-smoking-act-of-1970.

Bucher, Tamara, Clare Collins, Megan E. Rollo, Tracy A. McCaffrey, Nienke de Vlieger, Daphne van der Bend, Helen Truby, and Federico J. A. Perez-Cueto. "Nudging Consumers Towards Healthier Choices: A Systematic Review of Positional Influences on Food Choice." *British Journal of Nutrition* 115, no. 12 (2016): 2252–63. https://doi.org/10.1017/s0007114516001653.

Burns, Justine, Brendan Maughan-Brown, and Âurea Mouzinho. "Washing with Hope: Evidence from a Hand-Washing Pilot Study Among Children in South Africa." *BMC Public Health* 18 (2018): 709. https://doi.org/10.1186/s12889-018-5573-8.

Burns, Mark J. "Success Is Not an Accident: What Sports Business Millennials Can Learn from NBA MVP Stephen Curry." *Forbes*. June 13, 2015. https://www.forbes.com/sites/markjburns/2015/06/13/success-is-not-an-accident-what-sports-business-millennials-can-learn-from-nba-mvp-stephen-curry-2/#62c34b3d15fb.

Cantor, Jonathan, Alejandro Torres, Courtney Abrams, and Brian Elbel. "Five Years Later: Awareness of New York City's Calorie Labels Declined, with No Changes in Calories Purchased." *Health Affairs* 34, no. 11 (2015): 1893–1900. https://doi.org/10.1377/hlthaff.2015.0623.

Carli, Lorraine. "NFPA Encourages Testing Smoke Alarms as Daylight Saving Time Begins." National Fire Protection Association. March 6, 2014. https://www.nfpa.org/News-and-Research/News-and-media/Press-Room/News-releases/2014/NFPA-encourages-testing-smoke-alarms-as-Daylight-Saving-Time-begins.

Carrell, Scott E., Mark Hoekstra, and James E. West. "Is Poor Fitness Contagious? Evidence from Randomly Assigned Friends." *Journal of Public Economics* 95, nos. 7–8 (2011): 657–63. www.nber.org/papers/w16518.

Casagrande, Sarah Stark, Youfa Wang, Cheryl Anderson, and Tiffany L. Gary. "Have Americans Increased Their Fruit and Vegetable Intake? The Trends Between 1988 and 2002." *American Journal of Preventive Medicine* 32, no. 4 (2007): 257–63. https://doi.org/10.1016/j.amepre.2006.12.002.

Caton, Samantha J., Sara M. Ahern, Eloise Remy, Sophie Nicklaus, Pam Blundell, and Marion M. Hetherington. "Repetition Counts: Repeated Exposure Increases Intake of a Novel Vegetable in UK Pre-school Children Compared to Flavour–Flavour and Flavour–Nutrient Learning." *British Journal of Nutrition* 109, no. 11 (2013): 2089–97. https://doi.org/10.1017/s0007114512004126.

Centers for Disease Control and Prevention. "2014 State Indicator Report on Physical Activity." Atlanta, GA: U.S. Department of Health and Human Services, 2014. www.cdc.gov/physicalactivity/downloads/pa_state_indicator_report_2014.pdf.

———. "Burden of Tobacco Use in the U.S." Last modified April 23, 2018. https://www.cdc.gov/tobacco/campaign/tips/resources/data/cigarette-smoking-in-united-states.html.

———. "Cigarette Smoking and Tobacco Use Among People of Low Socioeconomic Status." Last modified August 21, 2018. https://www.cdc.gov/tobacco/disparities/low-ses/index.htm.

———. "Map of Current Cigarette Use Among Adults." September 19, 2017. https://www.cdc.gov/statesystem/cigaretteuseadult.html.

———. "Map of Excise Tax Rates on Cigarettes." January 2, 2018. https://www.cdc.gov/statesystem/excisetax.html.

———. "Quitting Smoking Among Adults—United States, 2000–2015." January 6, 2017. https://www.cdc.gov/mmwr/volumes/65/wr/mm6552a1.htm?s_cid=mm6552a1_w.

———. "Quitting Smoking Among Adults—United States, 2000–2015: Highlights." January 6, 2017. https://www.cdc.gov/tobacco/data_statistics/mmwrs/byyear/2017/mm6552a1/highlights.htm.

———. "State and Local Comprehensive Smoke-Free Laws for Worksites, Restaurants, and Bars—United States, 2015." Last modified August 24, 2017. https://www.cdc.gov/mmwr/volumes/65/wr/mm6524a4.htm.

———. "Tobacco-Related Mortality." May 15, 2017. https://www.cdc.gov/tobacco/data_statistics/fact_sheets/health_effects/tobacco_related_mortality/index.htm.

Chaiton, Michael, Lori Diemert, Joanna E. Cohen, Susan J. Bondy, Peter Selby, Anne Philipneri, and Robert Schwartz. "Estimating the Number of Quit Attempts It Takes to Quit Smoking Successfully in a Longitudinal Cohort of Smokers." *BMJ Open* 6, no. 6 (2016): e011045. https://doi.org/10.1136/bmjopen-2016-011045.

Chalabi, Mona. "How Many Times Does the Average Person Move?" *FiveThirtyEight*. January 29, 2015. https://fivethirtyeight.com/features/how-many-times-the-average-person-moves/.

Chandon, Pierre. "How Package Design and Packaged-Based Marketing Claims Lead to Overeating." *Applied Economic Perspectives and Policy* 35, no. 1 (2013): 7–31. https://doi.org/10.1093/aepp/pps028.

Colchero, M. Arantxa, Juan Rivera-Dommarco, Barry M. Popkin, and Shu Wen Ng. "In Mexico, Evidence of Sustained Consumer Response Two Years After Implementing a Sugar-Sweetened Beverage Tax." *Health Affairs* 36, no. 3 (2017): 564–71. https://doi.org/10.1377/hlthaff.2016.1231.

Cooper, Adam. "Electric Company Smart Meter Deployments: Foundation for a Smart Grid." Institute for Electric Innovation. December 2017. http://www.edisonfoundation.net/iei/publications/Documents/IEI_Smart%20Meter%20Report%202017_FINAL.pdf.

Crandall, Christian, and Monica Biernat. "The Ideology of Anti-fat Attitudes." *Journal of Applied Social Psychology* 20, no. 3 (1990): 227–43. https://doi.org/10.1111/j.1559-1816.1990.tb00408.x.

Crits-Christoph, Paul, Lynne Siqueland, Jack Blaine, Arlene Frank, Lester Luborsky, Lisa S. Onken, Larry R. Muenz, et al. "Psychosocial Treatments for Cocaine Dependence: National Institute on Drug Abuse Collaborative Cocaine Treatment Study." *Archives of General Psychiatry* 56, no. 6 (1999): 493–502. https://doi.org/10.1001/archpsyc.56.6.493.

Cruwys, Tegan, Kirsten E. Bevelander, and Roel C. J. Hermans. "Social Modeling of Eating: A Review of When and Why Social Influence Affects Food Intake and Choice." *Appetite* 86 (2015): 3–18. https://doi.org/10.1016/j.appet.2014.08.035.

Csikszentmihalyi, Mihaly. *Flow: The Psychology of Optimal Experience.* New York: Harper Perennial, 1996.

Danner, Unna N., Henk Aarts, and Nanne K. de Vries. "Habit vs. Intention in the Prediction of Future Behaviour: The Role of Frequency, Context Stability and Mental Accessibility of Past Behaviour." *British Journal of Social Psychology* 47, no. 2 (2008): 245–65. https://doi.org/10.1348/014466607x230876.

Deater-Deckard, Kirby, Michael D. Sewell, Stephen A. Petrill, and Lee A. Thompson. "Maternal Working Memory and Reactive Negativity in Parenting." *Psychological Science* 21, no. 1 (2010): 75–79. https://doi.org/10.1177/0956797609354073.

DeFulio, Anthony, and Kenneth Silverman. "Employment-Based Abstinence Reinforcement as a Maintenance Intervention for the Treatment of Cocaine Dependence: Post-intervention Outcomes." *Addiction* 106, no. 5 (2011): 960–67. https://doi.org/10.1111/j.1360-0443.2011.03364.x.

De Ridder, Denise T. D., Gerty Lensvelt-Mulders, Catrin Finkenauer, F. Marijn Stok, and Roy F. Baumeister. "Taking Stock of Self-Control: A Meta-Analysis of How Trait Self-Control Relates to a Wide Range of Behaviors." *Personality and Social Psychology Review* 16, no. 1 (2012): 76–99. https://doi.org/10.1177/1088868311418749.

DeRusso, Alicia, David Fan, Jay Gupta, Oksana Shelest, Rui M. Costa, and Henry H. Yin. "Instrumental Uncertainty as a Determinant of Behavior Under Interval Schedules of Reinforcement." *Frontiers in Integrative Neuroscience* 4 (2010). https://doi.org/10.3389/fnint.2010.00017.

DeSilver, Drew. "Perceptions and Realities of Recycling Vary Widely from Place to Place." Pew Research Center. October 7, 2016. http://www.pewresearch.org/fact-tank/2016/10/07/perceptions-and-realities-of-recycling-vary-widely-from-place-to-place.

Dettmers, Jan. "How Extended Work Availability Affects Well-Being: The Mediating Roles of Psychological Detachment and Work-Family Conflict." *Work and Stress* 31, no. 1 (2017): 24–41. https://doi.org/10.1080/02678373.2017.1298164.

Dettmers, Jan, Tim Vahle-Hinz, Eva Bamberg, Niklas Friedrich, and Monika Keller. "Extended Work Availability and Its Relation with Start-of-Day Mood and Cortisol." *Journal of Occupational Health Psychology* 21, no. 1 (2016): 105–18. http://doi.org/10.1037/a0039602.

Dickinson, Anthony, and Lawrence Weiskrantz. "Actions and Habits: The Development of Behavioural Autonomy." *Philosophical Transactions of the Royal Society of London. B: Biological Sciences* 308, no. 1135 (1985): 67–78. https://doi.org/10.1098/rstb.1985.0010.

Dieu-Hang, To, R. Quentin Grafton, Roberto Martínez-Espiñeira, and Maria Garcia-Valiñas. "Household Adoption of Energy and Water-Efficient Appliances: An Analysis of Attitudes, Labelling and Complementary Green Behaviours in Selected OECD Countries." *Journal of Environmental Management* 197 (2017): 140–50. https://doi.org/10.1016/j.jenvman.2017.03.070.

Diliberti, Nicole, Peter L. Bordi, Martha T. Conklin, Liane S. Roe, and Barbara J. Rolls. "Increased Portion Size Leads to Increased Energy Intake in a Restaurant Meal." *Obesity Research* 12, no. 3 (2004): 562–68. https://doi.org/10.1038/oby.2004.64.

Doll, Richard, and Richard Peto. "The Causes of Cancer: Quantitative Estimates of Avoidable Risks of Cancer in the United States Today." *JNCI: Journal of the National Cancer Institute* 66, no. 6 (1981): 1192–308. https://doi.org/10.1093/jnci/66.6.1192.

Duckworth, Angela L., Tamar Szabó Gendler, and James J. Gross. "Situational Strategies for Self-Control." *Perspectives on Psychological Science* 11, no. 1 (2016): 35–55. https://doi.org/10.1177/1745691615623247.

Duckworth, Angela L., Rachel E. White, Alyssa J. Matteucci, Annie Shearer, and James J. Gross. "A Stitch in Time: Strategic Self-Control in High School and College Students." *Journal of Educational Psychology* 108, no. 3 (2016): 329–41. https://doi.org/10.1037/edu0000062.

Dunning, Thad, Felipe Monestier, Rafael Pineiro, Fernando Rosenblatt, and Guadalupe Tunón. "Is Paying Taxes Habit Forming? Experimental Evidence from Uruguay." Paper presented at the University of California, Berkeley, 2017. http://www.thaddunning.com/wp-content/uploads/2017/09/Dunning-et-al_Habit_2017.pdf.

Durant, Will. *The Story of Philosophy: The Lives and Opinions of the World's Greatest Philosophers.* 1926. Reprint, New York: Pocket Books, 1954.

Eadicicco, Lisa. "Americans Check Their Phones 8 Billion Times a Day." *Time.* December 15, 2015. http://time.com/4147614/smartphone-usage-us-2015.

Ell, Kellie. "Video Game Industry Is Booming with Continued Revenue." CNBC. July 18, 2018. https://www.cnbc.com/2018/07/18/video-game-industry-is-booming-with-continued-revenue.html.

Ent, Michael R., Roy F. Baumeister, and Dianne M. Tice. "Trait Self-Control and the Avoidance of Temptation." *Personality and Individual Differences* 74 (2015): 12–15. https://doi.org/10.1016/j.paid.2014.09.031.

Evans, Alexandra E., Rose Jennings, Andrew W. Smiley, Jose L. Medina, Shreela V. Sharma, Ronda Rutledge, Melissa H. Stigler, and Deanna M. Hoelscher. "Introduction of Farm Stands in Low-Income Communities Increases Fruit and Vegetable Among Community Residents." *Health and Place* 18, no. 5 (2012): 1137–43. https://doi.org/10.1016/j.healthplace.2012.04.007.

Evans, Jonathan St. B. T., and Keith E. Stanovich. "Dual-Process Theories of Higher Cognition: Advancing the Debate." *Perspectives on Psychological Science* 8, no. 3 (2013): 223–41. https://doi.org/10.1177/1745691612460685.

Everitt, Barry J., and Trevor W. Robbins. "Drug Addiction: Updating Actions to Habits to Compulsions Ten Years On." *Annual Review of Psychology* 67, no. 1 (2016): 23–50. https://doi.org/10.1146/annurev-psych-122414-033457.

Festinger, Leon, Stanley Schachter, and Kurt Back. *Social Pressures in Informal Groups: A Study of Human Factors in Housing.* New York: Harper, 1950.

Finkel, Eli J., and W. Keith Campbell. "Self-Control and Accommodation in Close Relationships: An Interdependence Analysis." *Journal of Personality and Social Psychology* 81, no. 2 (2001): 263–77. https://doi.org/10.1037//0022-3514.81.2.263.

Florida, Richard. "The Geography of Car Deaths in America." *CityLab.* October 15, 2015. http://www.citylab.com/commute/2015/10/the-geography-of-car-deaths-in-america/410494.

Follingstad, Diane R., and Maryanne Edmundson. "Is Psychological Abuse Reciprocal in Intimate Relationships? Data from a National Sample of American Adults." *Journal of Family Violence* 25, no. 5 (2010): 495–508. doi:10.1007/s10896-010-9311-y.

Foster, Sarah, Georgina Trapp, Paula Hooper, Wendy H. Oddy, Lisa Wood, and Matthew Knuiman. "Liquor Landscapes: Does Access to Alcohol Outlets Influence Alcohol Consumption in Young Adults?" *Health and Place* 45 (2017): 17–23. https://doi.org/10.1016/j.healthplace.2017.02.008.

Frey, Erin, and Todd Rogers. "Persistence: How Treatment Effects Persist After Interventions Stop." *Policy Insights from the Behavioral and Brain Sciences* 1, no. 1 (2014): 172–79. https://doi.org/10.1177/2372732214550405.

Fujiwara, Thomas, Kyle Meng, and Tom Vogl. "Habit Formation in Voting: Evidence from Rainy Elections." *American Economic Journal: Applied Economics* 8, no. 4 (2016): 160–88. https://doi.org/10.1257/app.20140533.

Fulkerson, Jayne A., Mary Story, Alison Mellin, Nancy Leffert, Dianne Neumark-Sztainer, and Simone A. French. "Family Dinner Meal Frequency and Adolescent Development: Relationships with Developmental Assets and High-Risk Behaviors." *Journal of Adolescent Health* 39, no. 3 (2006): 337–45. https://doi.org/10.1016/j.jadohealth.2005.12.026.

Galaj, Ewa, Monica Manuszak, and Robert Ranaldi. "Environmental Enrichment as a Potential Intervention for Heroin Seeking." *Drug and Alcohol Dependence* 163 (2016): 195–201. https://doi.org/10.1016/j.drugalcdep.2016.04.016.

Galla, Brian M., and Angela L. Duckworth. "More Than Resisting Temptation: Beneficial Habits Mediate the Relationship Between Self-Control and Positive Life Outcomes." *Journal of Personality and Social Psychology* 109, no. 3 (2015): 508–25. https://doi.org/10.1037/pspp0000026.

Gardner, Benjamin, and Phillippa Lally. "Does Intrinsic Motivation Strengthen Physical Activity Habit? Modeling Relationships Between Self-Determination, Past Behaviour, and Habit Strength." *Journal of Behavioral Medicine* 36, no. 5 (2013): 488–97. https://doi.org/10.1007/s10865-012-9442-0.

Gates, Bill. *Business @ the Speed of Thought: Succeeding in the Digital Economy.* New York: Hachette, 1999.

Geertz, Clifford. *The Interpretation of Cultures.* New York: Basic Books, 1973.

Gillan, Claire M., A. Ross Otto, Elizabeth A. Phelps, and Nathaniel D. Daw. "Model-Based Learning Protects Against Forming Habits." *Cognitive, Affective, and Behavioral Neuroscience* 15, no. 3 (2015): 523–36. https://doi.org/10.3758/s13415-015-0347-6.

Gladwell, Malcolm. *Outliers: The Story of Success.* New York: Little, Brown, 2008.

Glantz, Stanton A. "Tobacco Taxes Are Not the Most Effective Tobacco Control Policy (As Actually Implemented)." UCSF Center for Tobacco Control Research and Education. January 11, 2014. https://tobacco.ucsf.edu/tobacco-taxes -are-not-most-effective-tobacco-control-policy-actually-implemented.

Gliklich, Emily, Rong Guo, and Regan W. Bergmark. "Texting While Driving: A Study of 1211 U.S. Adults with the Distracted Driving Survey." *Preventive Medicine Reports* 4 (2016): 486–89. https://doi.org/10.1016/j.pmedr.2016.09 .003.

Global Status Report on Road Safety 2018. Geneva: World Health Organization, 2018. https://www.who.int/violence_injury_prevention/road_safety_status/2018/en/.

Greenfield, Rebecca. "Workplace Wellness Programs Really Don't Work." Bloomberg. January 26, 2018. https://www.bloomberg.com/news/articles/2018-01-26 /workplace-wellness-programs-really-don-t-work.

Hadaway, Patricia F., Bruce K. Alexander, Robert B. Coambs, and Barry Beyerstein. "The Effect of Housing and Gender on Preference for Morphine-Sucrose Solutions in Rats." *Psychopharmacology* 66, no. 1 (1979): 87–91. https://doi.org /10.1007/bf00431995.

Hall, Matthew. "Bird Scooters Flying Around Town." *Santa Monica Daily Press.* September 26, 2017. http://smdp.com/bird-scooters-flying-around-town/162647.

Halpern, Daniel, and James E. Katz. "Texting's Consequences for Romantic Relationships: A Cross-Lagged Analysis Highlights Its Risks." *Computers in Human Behavior* 71 (2017): 386–94. https://doi.org/10.1016/j.chb.2017.01.051.

Hammons, Amber J., and Barbara H. Fiese. "Is Frequency of Shared Family Meals Related to the Nutritional Health of Children and Adolescents?" *Pediatrics* 127, no. 6 (2011): E1565–74. https://doi.org/10.1542/peds.2010-1440.

Harris, Mathew A., and Thomas Wolbers. "How Age-Related Strategy Switching Deficits Affect Wayfinding in Complex Environments." *Neurobiology of Aging* 35, no. 5 (2014): 1095–102. https://doi.org/10.1016/j.neurobiolaging.2013.10.086.

Harter, Jim. "Should Employers Ban Email After Work Hours?" Gallup. September 9, 2014. https://www.gallup.com/workplace/236519/employers-ban-email -work-hours.aspx.

Heatherton, Todd F., and Patricia A. Nichols. "Personal Accounts of Successful Versus Failed Attempts at Life Change." *Personality and Social Psychology Bulletin* 20, no. 6 (1994): 664–75. https://doi.org/10.1177/0146167294206005.

Heintzelman, Samantha J., and Laura A. King. "Routines and Meaning in Life." *Personality and Social Psychology Bulletin.* Published online September 18, 2018. https://doi.org/10.1177/0146167218795133.

Hirsch, Jana A., Ana V. Diez Roux, Kari A. Moore, Kelly R. Evenson, and Daniel A. Rodriguez. "Change in Walking and Body Mass Index Following Residential Relocation: The Multi-Ethnic Study of Atherosclerosis." *American Journal of Public Health* 104, no. 3 (2014): e49–e56. https://doi.org/10.2105/ajph.2013.301773.

Hobson, Nicholas M., Devin Bonk, and Michael Inzlicht. "Rituals Decrease the Neural Response to Performance Failure." *PeerJ* 5 (2017): e3363. https://doi .org/10.7717/peerj.3363.

Hoffman, Steven J., and Charlie Tan. "Overview of Systematic Reviews on the Health-Related Effects of Government Tobacco Control Policies." *BMC Public Health* 15, no. 1 (2015): 744. https://doi.org/10.1186/s12889-015-2041-6.

Hofford, Rebecca S., Jonathan J. Chow, Joshua S. Beckmann, and Michael T. Bardo. "Effects of Environmental Enrichment on Self-Administration of the Short-Acting Opioid Remifentanil in Male Rats." *Psychopharmacology* 234, nos. 23–24 (2017): 3499–506. https://doi.org/10.1007/s00213-017-4734-2.

Hofmann, Wilhelm, Roy F. Baumeister, Georg Förster, and Kathleen D. Vohs. "Everyday Temptations: An Experience Sampling Study of Desire, Conflict, and Self-Control." *Journal of Personality and Social Psychology* 102, no. 6 (2012): 1318–35, doi:10.1037/a0026545.

Hollands, Gareth J., Ian Shemilt, Theresa M. Marteau, Susan A. Jebb, Hannah B. Lewis, Yinghui Wei, Julian P. T. Higgins, and David Ogilvie. "Portion, Package or Tableware Size for Changing Selection and Consumption of Food, Alcohol and Tobacco." *Cochrane Database of Systematic Reviews* 9, no. CD011045 (2015): https://www.ncbi.nlm.nih.gov/pmc/articles/PMC4579823/.

Holmes, John G., and Susan D. Boon. "Developments in the Field of Close Relationships: Creating Foundations for Intervention Strategies." *Personality and Social Psychology Bulletin* 16, no. 1 (1990): 23–41. https://doi.org/10.1177/0146167290161003.

Howard-Jones, Paul A., Tim Jay, Alice Mason, and Harvey Jones. "Gamification of Learning Deactivates the Default Mode Network." *Frontiers in Psychology* 6 (2016). https://doi.org/10.3389/fpsyg.2015.01891.

Hui, Sam K., J. Jeffrey Inman, Yanliu Huang, and Jacob Suher. "The Effect of In-Store Travel Distance on Unplanned Spending: Applications to Mobile Promotion Strategies." *Journal of Marketing* 77, no. 2 (2013): 1–16. https://doi.org/10.1509/jm.11.0436.

Hunt, George M., and Nathan H. Azrin. "A Community-Reinforcement Approach to Alcoholism." *Behaviour Research and Therapy* 11, no. 1 (1973): 91–104. https://doi.org/10.1016/0005-7967(73)90072-7.

Hutson, Matthew. "Everyday Routines Make Life Feel More Meaningful." *Scientific American.* July 1, 2015. https://www.scientificamerican.com/article/everyday-routines-make-life-feel-more-meaningful/.

Itzchakov, Guy, Liad Uziel, and Wendy Wood. "When Attitudes and Habits Don't Correspond: Self-Control Depletion Increases Persuasion but Not Behavior." *Journal of Experimental Social Psychology* 75 (2018): 1–10. https://doi.org/10.1016/j.jesp.2017.10.011.

James, William. *Habit.* New York: Henry Holt, 1890.

———. *The Principles of Psychology*, vol. 1. New York: Henry Holt, 1890. Reprint, New York: Cosimo, 2007.

Jessoe, Katrina, and David Rapson. "Knowledge Is (Less) Power: Experimental Evidence from Residential Energy Use." *American Economic Review* 104, no. 4 (2014): 1417–38. https://doi.org/10.1257/aer.104.4.1417.

Ji, Mindy F., and Wendy Wood. "Purchase and Consumption Habits: Not Necessarily What You Intend." *Journal of Consumer Psychology* 17, no. 4 (2007): 261–76. https://doi.org/10.1016/S1057-7408(07)70037-2.

Jónsdóttir, María K., Steinunn Adólfsdóttir, Rúna Dögg Cortez, María Gunnarsdóttir, and Ágústa Hlín Gústafsdóttir. "A Diary Study of Action Slips in Healthy Individuals." *Clinical Neuropsychologist* 21, no. 6 (2007): 875–83. https://doi.org/10.1080/13854040701220044.

Jordan, Jewel. "Americans Moving at Historically Low Rates, Census Bureau Reports." United States Census Bureau. November 16, 2016. https://www.census .gov/newsroom/press-releases/2016/cb16-189.html.

Jost, John T., and David M. Amodio. "Political Ideology as Motivated Social Cognition: Behavioral and Neuroscientific Evidence." *Motivation and Emotion* 36, no. 1 (2012): 55–64. doi.10.1007/s11031-011-9260-7.

Judah, Gaby, Benjamin Gardner, and Robert Aunger. "Forming a Flossing Habit: An Exploratory Study of the Psychological Determinants of Habit Formation." *British Journal of Health Psychology* 18, no. 2 (2013): 338–53. https://doi.org/10 .1111/j.2044-8287.2012.02086.x.

Katz-Sidlow, Rachel J., Allison Ludwig, Scott Miller, and Robert Sidlow. "Smartphone Use During Inpatient Attending Rounds: Prevalence, Patterns and Potential for Distraction." *Journal of Hospital Medicine* 7, no. 8 (2012): 595–99. https://doi.org/10.1002/jhm.1950.

Kaushal, Navin, and Ryan E. Rhodes. "Exercise Habit Formation in New Gym Members: A Longitudinal Study." *Journal of Behavioral Medicine* 38, no. 4 (2015): 652–63. https://doi.org/10.1007/s10865-015-9640-7.

Keller, Carmen, Christina Hartmann, and Michael Siegrist. "The Association Between Dispositional Self-Control and Longitudinal Changes in Eating Behaviors, Diet Quality, and BMI." *Psychology and Health* 31, no. 11 (2016): 1311–27. https://doi.org/10.1080/08870446.2016.1204451.

Kessler, David A. *The End of Overeating: Taking Control of the Insatiable American Appetite*. Emmaus, PA: Rodale Books, 2009.

Khare, Adwait, and J. Jeffrey Inman. "Daily, Week-Part, and Holiday Patterns in Consumers' Caloric Intake." *Journal of Public Policy and Marketing* 28, no. 2 (2009): 234–52. https://doi.org/10.1509/jppm.28.2.234.

———. "Habitual Behavior in American Eating Patterns: The Role of Meal Occasions." *Journal of Consumer Research* 32, no. 4 (2006): 567–75. https://doi.org /10.1086/500487.

Kirchner, Thomas R., Jennifer Cantrell, Andrew Anesetti-Rothermel, Ollie Ganz, Donna M. Vallone, and David B. Abrams. "Geospatial Exposure to Point-of-Sale Tobacco: Real-Time Craving and Smoking-Cessation Outcomes." *American Journal of Preventive Medicine* 45, no. 4 (2013): 379–85. https://doi.org/10 .1016/j.amepre.2013.05.016.

Kiszko, Kamila M., Olivia D. Martinez, Courtney Abrams, and Brian Elbel. "The Influence of Calorie Labeling on Food Orders and Consumption: A Review of the Literature." *Journal of Community Health* 39, no. 6 (2014): 1248–69. https://doi.org/10.1007/s10900-014-9876-0.

Klein, Gary, Roberta Calderwood, and Anne Clinton-Cirocco. "Rapid Decision Making on the Fire Ground: The Original Study Plus a Postscript." *Journal of Cognitive Engineering and Decision Making* 4, no. 3 (2010): 186–209. https:// doi.org/10.1518/155534310X12844000801203.

Knowlton, Barbara J., Jennifer A. Mangels, and Larry R. Squire. "A Neostriatal Habit Learning System in Humans." *Science* 273, no. 5280 (1996): 1399–402. https://doi.org/10.1126/science.273.5280.1399.

Knowlton, Barbara J., and Tara K. Patterson. "Habit Formation and the Striatum." In *Behavioral Neuroscience of Learning and Memory*, eds. Robert E. Clark and

Stephen J. Martin, 275–95. Vol. 37 in *Current Topics in Behavioral Neurosciences*. Cham, Switzerland: Springer International, 2018. https://doi.org/10.1007/7854_2016_451.

Koehler, Derek J., Rebecca J. White, and Leslie K. John. "Good Intentions, Optimistic Self-Predictions, and Missed Opportunities." *Social Psychological and Personality Science* 2, no. 1 (2011): 90–96. https://doi.org/10.1177/1948550610375722.

Koob, George F., and Nora D. Volkow. "Neurobiology of Addiction: A Neurocircuitry Analysis." *Lancet Psychiatry* 3, no. 8 (2016): 760–73. https://doi.org/10.1016/S2215-0366(16)00104-8.

Korosec, Kirsten. "2016 Was the Deadliest Year on American Roads in Nearly a Decade." *Fortune*. February 15, 2017. http://fortune.com/2017/02/15/traffic-deadliest-year/.

Kullgren, Jeffrey T., Andrea B. Troxel, George Loewenstein, David A. Asch, Laurie A. Norton, Lisa Wesby, Yuanyuan Tao, et al. "Individual- Versus Group-Based Financial Incentives for Weight Loss: A Randomized, Controlled Trial." *Annals of Internal Medicine* 158, no. 7 (2013): 505–14. https://doi.org/10.7326/0003-4819-158-7-201304020-00002.

Kuzmarov, Jeremy. *The Myth of the Addicted Army: Vietnam and the Modern War on Drugs*. Amherst, MA: University of Massachusetts Press, 2009.

Labrecque, Jennifer S., Kristen Lee, and Wendy Wood. "Overthinking Habit." Manuscript under revision, University of Southern California, 2017.

Labrecque, Jennifer S., Wendy Wood, David T. Neal, and Nick Harrington. "Habit Slips: When Consumers Unintentionally Resist New Products." *Journal of the Academy of Marketing Science* 45, no. 1 (2017): 119–33. https://doi.org/10.1007/s11747-016-0482-9.

Lally, Phillippa, Cornelia H. M. van Jaarsveld, Henry W. W. Potts, and Jane Wardle. "How Are Habits Formed: Modelling Habit Formation in the Real World." *European Journal of Social Psychology* 40, no. 6 (2010): 998–1009. https://doi.org/10.1002/ejsp.674.

Larcom, Shaun, Ferdinand Rauch, and Tim Willems. "The Benefits of Forced Experimentation: Striking Evidence from the London Underground Network." *Quarterly Journal of Economics* 132, no. 4 (2017): 2019–55. https://doi.org/10.1093/qje/qjx020.

Legare, Cristine H., and André L. Souza. "Evaluating Ritual Efficacy: Evidence from the Supernatural." *Cognition* 124, no. 1 (2012): 1–15. https://doi.org/10.1016/j.cognition.2012.03.004.

Lewin, Kurt. "Frontiers in Group Dynamics: Concept, Method and Reality in Social Science; Social Equilibria and Social Change." *Human Relations* 1, no. 1 (1947): 5–41. https://doi.org/10.1177/001872674700100103.

Lewis, Zakkoyya H., Maria C. Swartz, and Elizabeth J. Lyons. "What's the Point? A Review of Reward Systems Implemented in Gamification Interventions." *Games for Health Journal* 5, no. 2 (2016): 93–99. https://doi.org/10.1089/g4h.2015.0078.

Lin, Pei-Ying, Wendy Wood, and John Monterosso. "Healthy Eating Habits Protect Against Temptations." *Appetite* 103 (2016): 432–40. https://doi.org/10.1016/j.appet.2015.11.011.

Litt, Mark D., Ronald M. Kadden, Elise Kabela-Cormier, and Nancy M. Petry. "Changing Network Support for Drinking: Network Support Project 2-Year

Follow-Up." *Journal of Consulting and Clinical Psychology* 77, no. 2 (2009): 229–42. https://doi.org/10.1037/a0015252.

Loewenstein, George, Cass R. Sunstein, and Russell Golman. "Disclosure: Psychology Changes Everything." *Annual Review of Economics* 6 (2014): 391–419. https://doi.org/10.1146/annurev-economics-080213-041341.

Lucas, Brian J., and Loran F. Nordgren. "People Underestimate the Value of Persistence for Creative Performance." *Journal of Personality and Social Psychology* 109, no. 2 (2015): 232–43. https://doi.org/10.1037/pspa0000030.

Lynley, Matthew. "Bird Has Officially Raised a Whopping $300M as the Scooter Wars Heat Up." *TechCrunch*. June 28, 2018. https://techcrunch.com/2018/06 /28/bird-has-officially-raised-a-whopping-300m-as-the-scooter-wars-heat-up.

Macnamara, Brooke N., David Z. Hambrick, and Frederick L. Oswald. "Deliberate Practice and Performance in Music, Games, Sports, Education, and Professions: A Meta-Analysis." *Psychological Science* 25, no. 8 (2014): 1608–18. https://doi.org/10.1177/0956797614535810.

Mader, Emily M., Brittany Lapin, Brianna J. Cameron, Thomas A. Carr, and Christopher P. Morley. "Update on Performance in Tobacco Control: A Longitudinal Analysis of the Impact of Tobacco Control Policy and the US Adult Smoking Rate, 2011–2013." *Journal of Public Health Management and Practice* 22, no. 5 (2016): E29–E35. https://doi.org/10.1097/phh.0000000000000358.

Maltz, Maxwell. *Psycho-Cybernetics*. New York: Pocket Books, 1989.

Mannor, Mike, Adam Wowak, Viva Ona Bartkus, and Luis R. Gomez-Mejia. "How Anxiety Affects CEO Decision Making." *Harvard Business Review*. July 19, 2016. https://hbr.org/2016/07/how-anxiety-affects-ceo-decision-making.

Mantzari, Eleni, Florian Vogt, Ian Shemilt, Yinghui Wei, Julian P. T. Higgins, and Theresa M. Marteau. "Personal Financial Incentives for Changing Habitual Health-Related Behaviors: A Systematic Review and Meta-Analysis." *Preventive Medicine* 75 (2015): 75–85. https://doi.org/10.1016/j.ypmed.2015.03.001.

March, James G. "Exploration and Exploitation in Organizational Learning." *Organization Science* 2, no. 1 (1991): 71–87. https://www.jstor.org/stable/2634940.

Martin, Adam, Jenna Panter, Marc Suhrcke, and David Ogilvie. "Impact of Changes in Mode of Travel to Work on Changes in Body Mass Index: Evidence from the British Household Panel Survey." *Journal of Epidemiology and Community Health* 69, no. 8 (2015): 753–61. https://doi.org/10.1136/jech-2014-205211.

Mayer, Stefan, and Jan R. Landwehr. "Objective Measures of Design Typicality." *Design Studies* 54 (2018): 146–61. https://doi.org/10.1016/j.destud.2017.09 .004.

———. "Objective Measures of Design Typicality That Predict Aesthetic Liking, Fluency, and Car Sales." In *Advances in Consumer Research* 44. Duluth, MN: Association for Consumer Research, 2016: 556–57.

McCarthy, Justin. "In U.S., Smoking Rate Lowest in Utah, Highest in Kentucky." Gallup. March 13, 2014. http://www.gallup.com/poll/167771/smoking-rate -lowest-utah-highest-kentucky.aspx.

McDaniel, Brandon T., and Sarah M. Coyne. "'Technoference': The Interference of Technology in Couple Relationships and Implications for Women's Personal and Relational Well-Being." *Psychology of Popular Media Culture* 5, no. 1 (2016): 85–98, http://doi.org/10.1037/ppm0000065.

McKay, James R. "Making the Hard Work of Recovery More Attractive for Those with Substance Use Disorders." *Addiction* 112, no. 5 (2017): 751–57. https://doi.org/10.1111/add.13502.

McKinlay, John B. "A Case for Re-focusing Upstream: The Political Economy of Illness." In *Applying Behavioral Sciences to Cardiovascular Risk*, Proceedings of the American Heart Association Conference, Seattle, WA, June 17–19, 1974, edited by A. J. Enelow and J. B. Henderson. Washington, DC: American Heart Association, 1975.

Melnikoff, David E., and John A. Bargh. "The Mythical Number Two." *Trends in Cognitive Sciences* 22, no. 4 (2018): 280–93. https://doi.org/10.1016/j.tics.2018.02.001.

MetLife Foundation. "What America Thinks: MetLife Foundation Alzheimer's Survey." February 2011. https://www.metlife.com/assets/cao/foundation/alzheimers-2011.pdf.

Michimi, Akihiko, and Michael C. Wimberly. "Associations of Supermarket Accessibility with Obesity and Fruit and Vegetable Consumption in the Conterminous United States." *International Journal of Health Geographics* 9, no. 1 (2010): 49. https://doi.org/10.1186/1476-072x-9-49.

Miller, George A. "The Cognitive Revolution: A Historical Perspective." *Trends in Cognitive Sciences* 7, no. 3 (2003): 141–44. https://doi.org/10.1016/S1364-6613(03)00029-9.

Miller, George A., Eugene Galanter, and Karl H. Pribram. *Plans and the Structure of Behavior.* New York: Adams-Bannister-Cox, 1986.

Mischel, Walter, and Ebbe B. Ebbesen. "Attention in Delay of Gratification." *Journal of Personality and Social Psychology* 16, no. 2 (1970): 329–37. https://doi.org/10.1037/h0029815.

Mita, Theodore H., Marshall Dermer, and Jeffrey Knight. "Reversed Facial Images and the Mere-Exposure Hypothesis." *Journal of Personality and Social Psychology* 35, no. 8 (1977): 597–601. https://doi.org/10.1037//0022-3514.35.8.597.

Molloy, Gerard J., Heather Graham, and Hannah McGuinness. "Adherence to the Oral Contraceptive Pill: A Cross-Sectional Survey of Modifiable Behavioural Determinants." *BMC Public Health* 12 (2012). https://doi.org/10.1186/1471-2458-12-838.

Monterosso, John, and Wendy Wood. "Habits of Successful Rehabilitation." Unpublished data, University of Southern California, 2017.

Montoya, R. Matthew, Robert S. Horton, Jack L. Vevea, Martyna Citkowicz, and Elissa A. Lauber. "A Re-examination of the Mere Exposure Effect: The Influence of Repeated Exposure on Recognition, Familiarity, and Liking." *Psychological Bulletin* 143, no. 5 (2017): 459–98. https://doi.org/10.1037/bul0000085.

Mooney, Chris. "Why 50 Million Smart Meters Still Haven't Fixed America's Energy Habits." *The Washington Post*, January 29, 2015. https://www.washingtonpost.com/news/energy-environment/wp/2015/01/29/americans-are-this-close-to-finally-understanding-their-electricity-bills.

Moore, Latetia V., and Frances E. Thompson. "Adults Meeting Fruit and Vegetable Intake Recommendations—United States 2013." *Morbidity and Mortality Weekly Report* 64, no. 26 (2015): 709–13. Washington, D.C.: Centers for Disease Control and Prevention, July 10, 2015. https://www.cdc.gov/mmwr/preview/mmwrhtml/mm6426a1.htm.

Morley, Christopher P., and Morgan A. Pratte. "State-Level Tobacco Control and Adult Smoking Rate in the United States: An Ecological Analysis of Structural Factors." *Journal of Public Health Management and Practice* 19, no. 6 (2013): E20–E27. https://doi.org/10.1097/PHH.0b013e31828000de.

Morris, Benjamin. "Stephen Curry Is the Revolution." *FiveThirtyEight*. December 3, 2015. http://fivethirtyeight.com/features/stephen-curry-is-the-revolution.

Mosley, Michael. "Five-A-Day Campaign: A Partial Success." BBC News. January 3, 2013. http://www.bbc.com/news/health-20858809.

Nasar, Jack L., and Derek Troyer. "Pedestrian Injuries Due to Mobile Phone Use in Public Places." *Accident Analysis and Prevention* 57 (2013): 91–95. https://doi.org/10.1016/j.aap.2013.03.021.

NatCen Social Research. *Health Survey for England 2017.* London: NHS Digital, 2018. https://files.digital.nhs.uk/5B/B1297D/HSE%20report%20summary.pdf.

National Association of City Transportation Officials. *Equitable Bike Share Means Building Better Places for People to Ride.* July 2016. https://nacto.org/wp-content/uploads/2016/07/NACTO_Equitable_Bikeshare_Means_Bike_Lanes.pdf.

National Heart, Lung, and Blood Institute. "Portion Distortion." Last modified on April 1, 2015. https://www.nhlbi.nih.gov/health/educational/wecan/eat-right/portion-distortion.htm.

National Institute on Drug Abuse. "Drugs, Brains, and Behavior: The Science of Addiction." Last modified July 2018. https://www.drugabuse.gov/publications/drugs-brains-behavior-science-addiction/drug-abuse-addiction.

———. "Drugs, Brains, and Behavior: The Science of Addiction: Treatment and Recovery." July 2014. https://www.drugabuse.gov/publications/drugs-brains-behavior-science-addiction/treatment-recovery.

National Safety Council Injury Facts. "Odds of Dying." 2016. https://injuryfacts.nsc.org/all-injuries/preventable-death-overview/odds-of-dying.

Neal, David T., Jelena Vujcic, Orlando Hernandez, and Wendy Wood. *The Science of Habit: Creating Disruptive and Sticky Behavior Change in Handwashing Behavior.* Washington, DC: USAID/WASHplus Project, 2015.

Neal, David T., Wendy Wood, and Aimee Drolet. "How Do People Adhere to Goals When Willpower Is Low? The Profits (and Pitfalls) of Strong Habits." *Journal of Personality and Social Psychology* 104, no. 6 (2013): 959–75. https://doi.org/10.1037/a0032626.

Neal, David T., Wendy Wood, Jennifer S. Labrecque, and Phillippa Lally. "How Do Habits Guide Behavior? Perceived and Actual Triggers of Habits in Daily Life." *Journal of Experimental Social Psychology* 48, no. 2 (2012): 492–98. https://doi.org/10.1016/j.jesp.2011.10.011.

Neal, David T., Wendy Wood, Mengju Wu, and David Kurlander. "The Pull of the Past: When Do Habits Persist Despite Conflict with Motives?" *Personality and Social Psychology Bulletin* 37, no. 11 (2011): 1428–37. http://doi.org/10.1177/0146167211419863.

Newport, Frank. "Email Outside of Working Hours Not a Burden to U.S. Workers." Gallup. May 10, 2017. https://news.gallup.com/poll/210074/email-outside-working-hours-not-burden-workers.aspx.

Nisbett, Richard E., and Timothy D. Wilson. "Telling More Than We Can Know: Verbal Reports on Mental Processes." *Psychological Review* 84, no. 3 (1977): 231–59. https://doi.org/10.1037/0033-295X.84.3.231.

NORC at the University of Chicago. "New Insights into Americans' Perceptions and Misperceptions of Obesity Treatments, and the Struggles Many Face." October 2016. http://www.norc.org/PDFs/ASMBS%20Obesity/ASMBS%20 NORC%20Obesity%20Poll_Brief%20B%20REV010917.pdf.

———. "The ASMBS and NORC Survey on Obesity in America." Accessed March 10, 2018. http://www.norc.org/Research/Projects/Pages/the-asmbsnorc -obesity-poll.aspx.

Norton, Michael I., and Francesca Gino. "Rituals Alleviate Grieving for Loved Ones, Lovers, and Lotteries." *Journal of Experimental Psychology: General* 143, no. 1 (2014): 266–72. https://doi.org/10.1037/a0031772.

Nutt, David J., Anne Lingford-Hughes, David Erritzoe, and Paul R. A. Stokes. "The Dopamine Theory of Addiction: 40 Years of Highs and Lows." *Nature Reviews Neuroscience* 16, no. 5 (2015): 305–12. https://doi.org/10.1038/nrn3939.

NYC DOT. *Cycling in the City: Cycling Trends in NYC.* 2018. http://www.nyc.gov /html/dot/downloads/pdf/cycling-in-the-city.pdf.

Obermeier, Christian, Sonja A. Kotz, Sarah Jessen, Tim Raettig, Martin von Koppenfels, and Winfried Menninghaus. "Aesthetic Appreciation of Poetry Correlates with Ease of Processing in Event-Related Potentials." *Cognitive, Affective, and Behavioral Neuroscience* 16, no. 2 (2016): 362–73. https://doi.org /10.3758/s13415-015-0396-x.

Orbell, Sheina, and Bas Verplanken. "The Automatic Component of Habit in Health Behavior: Habit as Cue-Contingent Automaticity." *Health Psychology* 29, no. 4 (2010): 374–83. https://doi.org/10.1037/a0019596.

Ozcelik, Erol, Nergiz Ercil Cagiltay, and Nese Sahin Ozcelik. "The Effect of Uncertainty on Learning in Game-Like Environments." *Computers and Education* 67 (2013): 12–20. https://doi.org/10.1016/j.compedu.2013.02.009.

Park-Lee, Eunice, Rachel N. Lipari, Sarra L. Hedden, Larry A. Kroutil, and Jeremy D. Porter. *Receipt of Services for Substance Use and Mental Health Issues Among Adults: Results from the 2016 National Survey on Drug Use and Health.* Rockville, MD: SAMHSA: NSDUH Data Review, September 2017.

Partners Studio. "4 Reasons Why Over 50% Car Crashes Happen Closer to Home." *HuffPost.* December 14, 2017. https://www.huffingtonpost.co.za/2017/12/14/4 -reasons-why-over-50-car-crashes-happen-closer-to-home_a_23307197.

Patterson, Tara K., and Barbara J. Knowlton. "Subregional Specificity in Human Striatal Habit Learning: A Meta-Analytic Review of the fMRI Literature." *Current Opinion in Behavioral Sciences* 20 (2018): 75–82. https://doi.org/10 .1016/j.cobeha.2017.10.005.

Payesko, Jenna. "FDA Approves Lofexidine Hydrochloride, First Non-opioid Treatment for Management of Opioid Withdrawal Symptoms in Adults." *Med Magazine,* May 16, 2018. https://www.mdmag.com/medical-news/fda-approves -lofexidine-hydrochloride-first-nonopioid-treatment-for-management-of-opioid -withdrawal-symptoms-in-adults.

Phillips, L. Alison, Howard Leventhal, and Elaine A. Leventhal. "Assessing Theoretical Predictors of Long-Term Medication Adherence: Patients' Treatment-

Related Beliefs, Experiential Feedback and Habit Development." *Psychology and Health* 28, no. 10 (2013): 1135–51. https://doi.org/10.1080/08870446.2013.793798.

Pollan, Michael. "The Way We Live Now: 10-12-03; The (Agri)Cultural Contradictions of Obesity." *The New York Times Magazine*, October 12, 2003. http://www.nytimes.com/2003/10/12/magazine/the-way-we-live-now-10-12-03-the-agri-cultural-contradictions-of-obesity.html.

Posavac, Steven S., Frank R. Kardes, and J. Joško Brakus. "Focus Induced Tunnel Vision in Managerial Judgment and Decision Making: The Peril and the Antidote." *Organizational Behavior and Human Decision Processes* 113, no. 2 (2010): 102–11. https://doi.org/10.1016/j.obhdp.2010.07.002.

Privitera, Gregory J., and Faris M. Zuraikat. "Proximity of Foods in a Competitive Food Environment Influences Consumption of a Low Calorie and a High Calorie Food." *Appetite* 76 (2014): 175–79. https://doi.org/10.1016/j.appet.2014.02.004.

Pronin, Emily, and Matthew B. Kugler. "People Believe They Have More Free Will Than Others." *Proceedings of the National Academy of Sciences* 107, no. 52 (2010): 22469–74. https://doi.org/10.1073/pnas.1012046108.

Quinn, Jeffrey M., Anthony Pascoe, Wendy Wood, and David T. Neal. "Can't Control Yourself? Monitor Those Bad Habits." *Personality and Social Psychology Bulletin* 36, no. 4 (2010): 499–511. https://doi.org/10.1177/0146167209360665.

Quinn, Jeffrey M., and Wendy Wood. "Habits Across the Lifespan." Unpublished manuscript, Duke University, 2005.

Ravaisson, Félix. *Of Habit*. Translated by Clare Carlisle and Mark Sinclair. 1838. Reprint, London: Continuum, 2008.

Reason, James, and Deborah Lucas. "Absent-Mindedness in Shops: Its Incidence, Correlates and Consequences." *British Journal of Clinical Psychology* 23, no. 2 (1984): 121–31. https://doi.org/10.1111/j.2044-8260.1984.tb00635.x.

Reber, Rolf, Norbert Schwarz, and Piotr Winkielman. "Processing Fluency and Aesthetic Pleasure: Is Beauty in the Perceiver's Processing Experience?" *Personality and Social Psychology Review* 8, no. 4 (2004): 364–82. https://doi.org/10.1207/s15327957pspr0804_3.

Reddit. "I'm Bill Gates, Co-chair of the Bill and Melinda Gates Foundation. Ask Me Anything." Accessed May 14, 2018. https://www.reddit.com/r/IAmA/comments/49jkhn/im_bill_gates_cochair_of_the_bill_melinda_gates/.

Redgrave, Peter, Manuel Rodriguez, Yoland Smith, Maria C. Rodriguez-Oroz, Stephane Lehericy, Hagai Bergman, Yves Agid, Mahlon R. DeLong, and José A. Obeso. "Goal-Directed and Habitual Control in the Basal Ganglia: Implications for Parkinson's Disease." *Nature Reviews Neuroscience* 11, no. 11 (2010): 760–72. https://doi.org/10.1038/nrn2915.

Roberts, James A., and Meredith E. David. "My Life Has Become a Major Distraction from My Cell Phone: Partner Phubbing and Relationship Satisfaction Among Romantic Partners." *Computers in Human Behavior* 54 (2016): 134–41. https://doi.org/10.1016/j.chb.2015.07.058.

Robins, Lee N. "Vietnam Veterans' Rapid Recovery from Heroin Addiction: A Fluke or Normal Expectation?" *Addiction* 88, no. 8 (1993): 1041–54. https://doi.org/10.1111/j.1360-0443.1993.tb02123.x.

Robins, Lee N., Darlene H. Davis, and Donald W. Goodwin. "Drug Use by US Army Enlisted Men in Vietnam: A Follow-Up on Their Return Home." *American Journal of Epidemiology* 99, no. 4 (1974): 235–49. https://doi.org/10.1093/oxfordjournals.aje.a121608.

Robins, Lee N., John E. Helzer, Michie Hesselbrock, and Eric Wish. "Vietnam Veterans Three Years After Vietnam: How Our Study Changed Our View of Heroin." *American Journal on Addictions* 19, no. 3 (2010): 203–11. https://doi.org/10.1111/j.1521-0391.2010.00046.x.

Robinson, Paul L., Fred Dominguez, Senait Teklehaimanot, Martin Lee, Arleen Brown, Michael Goodchild, and Darryl B. Hood. "Does Distance Decay Modelling of Supermarket Accessibility Predict Fruit and Vegetable Intake by Individuals in a Large Metropolitan Area?" *Journal of Health Care for the Poor and Underserved* 24, no. 1A (2013): 172–85. https://doi.org/10.1353/hpu.2013.0049.

Robinson, Thomas N., Dina L. G. Borzekowski, Donna M. Matheson, and Helena C. Kraemer. "Effects of Fast Food Branding on Young Children's Taste Preferences." *Archives of Pediatrics and Adolescent Medicine* 161, no. 8 (2007): 792–97. https://doi.org/10.1001/archpedi.161.8.792.

Rogers, Bryan L., James M. Vardaman, David G. Allen, Ivan S. Muslin, and Meagan Brock Baskin. "Turning Up by Turning Over: The Change of Scenery Effect in Major League Baseball." *Journal of Business and Psychology* 32, no. 5 (2017): 547–60. https://doi.org/10.1007/s10869-016-9468-3.

Rolls, Barbara J., Liane S. Roe, and Jennifer S. Meengs. "The Effect of Large Portion Sizes on Energy Intake Is Sustained for 11 Days." *Obesity* 15, no. 6 (2007): 1535–43. https://doi.org/10.1038/oby.2007.182.

Rosengren, John. "How Casinos Enable Gambling Addicts." *The Atlantic*, December 2016. https://www.theatlantic.com/magazine/archive/2016/12/losing-it-all/505814/.

Ross, Lee D., Teresa M. Amabile, and Julia L. Steinmetz. "Social Roles, Social Control, and Biases in Social-Perception Processes." *Journal of Personality and Social Psychology* 35, no. 7 (1977): 485–94. https://doi.org/10.1037/0022-3514.35.7.485.

Rothman, Michael. "Stephen and Ayesha Curry: Inside Our Whirlwind Life." ABC News. Accessed May 18, 2018. https://abcnews.go.com/Entertainment/fullpage/stephen-ayesha-curry-inside-whirlwind-life-34207323.

Runnemark, Emma, Jonas Hedman, and Xiao Xiao. "Do Consumers Pay More Using Debit Cards Than Cash?" *Electronic Commerce Research and Applications* 14, no. 5 (2015): 285–91. https://doi.org/10.1016/j.elerap.2015.03.002.

Ryan, Tom. "Older Shoppers Irritated by Supermarket Layout Changes." RetailWire, March 12, 2012. http://www.retailwire.com/discussion/older-shoppers-irritated-by-supermarket-layout-changes/.

Saad, Lydia. "Tobacco and Smoking." Gallup, August 15, 2002. http://www.gallup.com/poll/9910/tobacco-smoking.aspx.

———. "U.S. Smoking Rate Still Coming Down." Gallup. July 24, 2008. https://news.gallup.com/poll/109048/us-smoking-rate-still-coming-down.aspx.

Sanger-Katz, Margot. "The Decline of Big Soda." *The New York Times*, October 2, 2015. https://www.nytimes.com/2015/10/04/upshot/soda-industry-struggles-as-consumer-tastes-change.html.

———. "Yes, Soda Taxes Seem to Cut Soda Drinking." *The New York Times*, October 13, 2015. https://www.nytimes.com/2015/10/13/upshot/yes-soda-taxes-seem-to-cut-soda-drinking.html.

Scarboro, Morgan. "How High Are Cigarette Taxes in Your State?" Tax Foundation. May 10, 2017. https://taxfoundation.org/state-cigarette-taxes/.

Schippers, Michaéla C., and Paul A. M. van Lange. "The Psychological Benefits of Superstitious Rituals in Top Sport: A Study Among Top Sportspersons." *Journal of Applied Social Psychology* 36, no. 10 (2006): 2532–53. https://doi.org/10.1111/j.0021-9029.2006.00116.x.

Schlam, Tanya R., Nicole L. Wilson, Yuichi Shoda, Walter Mischel, and Ozlem Ayduk. "Preschoolers' Delay of Gratification Predicts Their Body Mass 30 Years Later." *Journal of Pediatrics* 162, no. 1 (2013): 90–93. https://doi.org/10.1016/j.jpeds.2012.06.049.

Schmidt, Susanne, and Martin Eisend. "Advertising Repetition: A Meta-Analysis on Effective Frequency in Advertising." *Journal of Advertising* 44, no. 4 (2015): 415–28. https://doi.org/10.1080/00913367.2015.1018460.

Schneider, Walter, and Richard M. Shiffrin. "Controlled and Automatic Human Information Processing: I. Detection, Search, and Attention." *Psychological Review* 84, no. 1 (1977): 1–66. https://doi.org/10.1037/0033-295X.84.1.1.

Schultz, Wolfram. "Dopamine Reward Prediction-Error Signalling: A Two-Component Response." *Nature Reviews Neuroscience* 17, no. 3 (2016): 183–95. https://doi.org/10.1038/nrn.2015.26.

———. "Dopamine Reward Prediction Error Coding." *Dialogues in Clinical Neuroscience* 18, no. 1 (2016): 23–32.

———. "Neuronal Reward and Decision Signals: From Theories to Data." *Physiological Reviews* 95, no. 3 (2015): 853–951. https://doi.org/10.1152/physrev.00023.2014.

Schwabe, Lars, and Oliver T. Wolf. "Stress and Multiple Memory Systems: From 'Thinking' to 'Doing.'" *Trends in Cognitive Sciences* 17, no. 2 (2013): 60–68. https://doi.org/10.1016/j.tics.2012.12.001.

———. "Stress Increases Behavioral Resistance to Extinction." *Psychoneuroendocrinology* 36, no. 9 (2011): 1287–93. https://doi.org/10.1016/j.psyneuen.2011.02.002.

Schwartz, Janet, Daniel Mochon, Lauren Wyper, Josiase Maroba, Deepak Patel, and Dan Ariely. "Healthier by Precommitment." *Psychological Science* 25, no. 2 (2014): 538–46. https://doi.org/10.1177/0956797613510950.

Schwarz, Hunter. "Where in the United States You Can't Purchase Alcohol." *The Washington Post*. September 2, 2014. https://www.washingtonpost.com/blogs/govbeat/wp/2014/09/02/where-in-the-united-states-you-cant-purchase-alcohol.

Sellman, Abigail, Justine Burns, and Brendan Maughan-Brown. "Handwashing Behaviour and Habit Formation in the Household: Evidence of Spillovers from a Pilot Randomised Evaluation in South Africa." SALDRU Working Paper Series, no. 226 (2018).

Sheeran, Paschal, Gaston Godin, Mark Conner, and Marc Germain. "Paradoxical Effects of Experience: Past Behavior Both Strengthens and Weakens the Intention-Behavior Relationship." *Journal of the Association for Consumer Research* 2, no. 3 (2017): 309–18. http://doi.org/10.1086/691216.

Shen, Luxi, Ayelet Fishbach, and Christopher K. Hsee. "The Motivating-Uncertainty Effect: Uncertainty Increases Resource Investment in the Process of Reward Pursuit." *Journal of Consumer Research* 41, no. 5 (2015): 1301–15. https://doi.org /10.1086/679418.

Shenhav, Amitai, Sebastian Musslick, Falk Lieder, Wouter Kool, Thomas L. Griffiths, Jonathan D. Cohen, and Matthew M. Botvinick. "Toward a Rational and Mechanistic Account of Mental Effort." *Annual Review of Neuroscience* 40 (2017): 99–124. https://doi.org/10.1146/annurev-neuro-072116-031526.

Shepherd, Lee, Ronan E. O'Carroll, and Eamonn Ferguson. "An International Comparison of Deceased and Living Organ Donation/Transplant Rates in Opt-In and Opt-Out Systems: A Panel Study." *BMC Medicine* 12, no. 1 (2014): 1–14. https://doi.org/10.1186/s12916-014-0131-4.

Shields, Grant S., Matthew A. Sazma, and Andrew P. Yonelinas. "The Effects of Acute Stress on Core Executive Functions: A Meta-Analysis and Comparison with Cortisol." *Neuroscience and Biobehavioral Reviews* 68 (2016): 651–68. https://doi.org/10.1016/j.neubiorev.2016.06.038.

Shiffrin, Richard M., and Walter Schneider. "Controlled and Automatic Human Information Processing: II. Perceptual Learning, Automatic Attending and a General Theory." *Psychological Review* 84, no. 2 (1977): 127–90. https://doi.org /10.1037/0033-295X.84.2.127.

Shindou, Tomomi, Mayumi Shindou, Sakurako Watanabe, and Jeff Wickens. "A Silent Eligibility Trace Enables Dopamine-Dependent Synaptic Plasticity for Reinforcement Learning in the Mouse Striatum." *European Journal of Neuroscience*. 2018: 1–11. https://doi.org/10.1111/ejn.13921.

Shoda, Yuichi, Walter Mischel, and Philip K. Peake. "Predicting Adolescent Cognitive and Self-Regulatory Competencies from Preschool Delay of Gratification: Identifying Diagnostic Conditions." *Developmental Psychology* 26, no. 6 (1990): 978–86. https://doi.org/10.1037/0012-1649.26.6.978.

Shrikant, Aditi. "11 Senior Citizens on the Best Products of the Past Century." *Vox*, December 11, 2018. https://www.vox.com/the-goods/2018/12/11/18116313/best -products-seniors-elderly-tide-samsung.

Shuster, Alvin M. "G.I. Heroin Addiction Epidemic in Vietnam." *The New York Times*. May 16, 1971. http://www.nytimes.com/1971/05/16/archives/gi-heroin -addiction-epidemic-in-vietnam-gi-heroin-addiction-is.html.

Silver, Lynn D., Shu Wen Ng, Suzanne Ryan-Ibarra, Lindsey Smith Taillie, Marta Induni, Donna R. Miles, Jennifer M. Poti, and Barry M. Popkin. "Changes in Prices, Sales, Consumer Spending, and Beverage Consumption One Year After a Tax on Sugar-Sweetened Beverages in Berkeley, California, US: A Before-and-After Study." *PLoS Medicine* 14, no. 4 (2017): e1002283. https://doi .org/10.1371/journal.pmed.1002283.

Silverman, Kenneth, Anthony DeFulio, and Sigurdur O. Sigurdsson. "Maintenance of Reinforcement to Address the Chronic Nature of Drug Addiction." *Preventive Medicine* 55 (2012): S46–S53. https://doi.org/10.1016/j.ypmed.2012.03.013.

Silverman, Kenneth, August F. Holtyn, and Reed Morrison. "The Therapeutic Utility of Employment in Treating Drug Addiction: Science to Application." *Translational Issues in Psychological Science* 2, no. 2 (2016): 203–12. https://doi.org /10.1037/tps0000061.

Sinclair, Susan E., Marcia Cooper, and Elizabeth D. Mansfield. "The Influence of Menu Labeling on Calories Selected or Consumed: A Systematic Review and Meta-Analysis." *Journal of the Academy of Nutrition and Dietetics* 114, no. 9 (2014): 1375–88. https://doi.org/10.1016/j.jand.2014.05.014.

Smith, Trevor, Edward Darling, and Bruce Searles. "2010 Survey on Cell Phone Use While Performing Cardiopulmonary Bypass." *Perfusion* 26, no. 5 (2011): 375–80. https://doi.org/10.1177/0267659111409969.

Snoek, Anke, Neil Levy, and Jeanette Kennett. "Strong-Willed but Not Successful: The Importance of Strategies in Recovery from Addiction." *Addictive Behaviors Reports* 4 (2016): 102–107. https://doi.org/10.1016/j.abrep.2016.09.002.

Spanos, Samantha, Lenny R. Vartanian, C. Peter Herman, and Janet Polivy. "Failure to Report Social Influences on Food Intake: Lack of Awareness or Motivated Denial?" *Health Psychology* 33, no. 12 (2014): 1487–94. https://doi.org/10.1037/hea0000008.

Spiegel, Alix. "What Vietnam Taught Us About Breaking Bad Habits." NPR. January 2, 2012. http://www.npr.org/sections/health-shots/2012/01/02/144431794/what-vietnam-taught-us-about-breaking-bad-habits.

Stables, Gloria, Jerianne Heimendinger, Mary Ann van Duyn, Linda Nebeling, Blossom Patterson, and Susan Berkowitz. "5 A Day Program Evaluation Research." In 5 *A Day for Better Health Program Monograph*, edited by Gloria Stables and Jerianne Heimendinger. Rockville, MD: MasiMax, 2001, 89–111.

Sternberg, Steve. "How Many Americans Floss Their Teeth?" *U.S. News and World Report*. May 2, 2016. https://www.usnews.com/news/articles/2016-05-02/how-many-americans-floss-their-teeth.

Stothart, Cary, Ainsley Mitchum, and Courtney Yehnert. "The Attentional Cost of Receiving a Cell Phone Notification." *Journal of Experimental Psychology: Human Perception and Performance* 41, no. 4 (2015): 893–97. http://doi.org/10.1037/xhp0000100.

Strömbäck, Camilla, Thérèse Lind, Kenny Skagerlund, Daniel Västfjäll, and Gustav Tinghög. "Does Self-Control Predict Financial Behavior and Financial Well-Being?" *Journal of Behavioral and Experimental Finance* 14 (2017): 30–38. https://doi.org/10.1016/j.jbef.2017.04.002.

Taillie, Lindsey Smith, Juan A. Rivera, Barry M. Popkin, and Carolina Batis. "Do High vs. Low Purchasers Respond Differently to a Nonessential Energy-Dense Food Tax? Two-Year Evaluation of Mexico's 8% Nonessential Food Tax." *Preventive Medicine* 105 (2017): S37–S42. https://doi.org/10.1016/j.ypmed.2017.07.009.

Tangney, June P., Roy F. Baumeister, and Angie Luzio Boone. "High Self-Control Predicts Good Adjustment, Less Pathology, Better Grades, and Interpersonal Success." *Journal of Personality* 72, no. 2 (2004). https://doi.org/10.1111/j.0022-3506.2004.00263.x.

Thaler, Richard H., and Cass R. Sunstein. *Nudge: Improving Decisions About Health, Wealth, and Happiness*. Updated edition. New York: Penguin, 2009.

Thiel, Kenneth J., Federico Sanabria, Nathan S. Pentkowski, and Janet L. Neisewander. "Anti-Craving Effects of Environmental Enrichment." *International Journal of Neuropsychopharmacology* 12, no. 9 (2009): 1151–56. https://doi.org/10.1017/S1461145709990472.

Thrailkill, Eric A., Sydney Trask, Pedro Vidal, José A. Alcalá, and Mark E. Bouton. "Stimulus Control of Actions and Habits: A Role for Reinforcer Predictability and Attention in the Development of Habitual Behavior." *Journal of Experimental Psychology: Animal Learning and Cognition* 44, no. 4 (2018): 370–84. https://doi.org/10.1037/xan0000188.

Tian, Allen Ding, Juliana Schroeder, Gerald Häubl, Jane L. Risen, Michael I. Norton, and Francesca Gino. "Enacting Rituals to Improve Self-Control." *Journal of Personality and Social Psychology* 114, no. 6 (2018): 851–76. https://doi.org/10.1037/pspa0000113.

Titchener, Edward Bradford. *A Text Book of Psychology*. Revised edition. New York: Macmillan, 1909.

"Tobacco: Data and Statistics." World Health Organization. Accessed February 16, 2019. http://www.euro.who.int/en/health-topics/disease-prevention/tobacco/data-and-statistics.

Tolman, Edward C. "Cognitive Maps in Rats and Men." *Psychological Review* 55, no. 4 (1948): 189–208. https://doi.org/10.1037/h0061626.

Tomek, Seven E., and M. Foster Olive. "Social Influences in Animal Models of Opiate Addiction." *International Review of Neurobiology* 140 (2018): 81–107. https://doi.org/10.1016/bs.irn.2018.07.004.

Umoh, Ruth. "Bill Gates Said He Had to Quit This Common Bad Habit Before He Became Successful." CNBC. March 16, 2018. https://www.cnbc.com/2018/03/16/bill-gates-quit-this-bad-habit-before-he-became-successful.html.

United States Department of Health and Human Services. *The Health Consequences of Smoking: 50 Years of Progress. A Report of the Surgeon General*. Atlanta, GA: U.S. Department of Health and Human Services, Centers for Disease Control and Prevention, National Center for Chronic Disease Prevention and Health Promotion, Office on Smoking and Health, 2014.

United States Department of Labor. "Employee Tenure Summary." Bureau of Labor Statistics. September 22, 2016. https://www.bls.gov/news.release/tenure.nr0.htm.

United States Public Health Service. *Smoking and Health: A Report of the Surgeon General: Appendix: Cigarette Smoking in the United States, 1950–1978*. United States Public Health Service, Office on Smoking and Health, 1979. https://profiles.nlm.nih.gov/ps/access/nnbcph.pdf.

VanDellen, Michelle R., James Y. Shah, N. Pontus Leander, Julie E. Delose, and Jerica X. Bornstein. "In Good Company: Managing Interpersonal Resources That Support Self-Regulation." *Personality and Social Psychology Bulletin* 41, no. 6 (2015): 869–82. https://doi.org/10.1177/0146167215580778.

Vangeli, Eleni, John Stapleton, Eline S. Smit, Ron Borland, and Robert West. "Predictors of Attempts to Stop Smoking and Their Success in Adult General Population Samples: A Systematic Review." *Addiction* 106, no. 12 (2011): 2110–21. https://doi.org/10.1111/j.1360-0443.2011.03565.x.

Vartanian, Lenny R., Samantha Spanos, C. Peter Herman, and Janet Polivy. "Conflicting Internal and External Eating Cues: Impact on Food Intake and Attributions." *Health Psychology* 36, no. 4 (2017): 365–69. https://doi.org/10.1037/hea0000447.

———. "Modeling of Food Intake: A Meta-Analytic Review." *Social Influence* 10, no. 3 (2015): 119–36. https://doi.org/10.1080/15534510.2015.1008037.

Verplanken, Bas, Henk Aarts, and Ad van Knippenberg. "Habit, Information Acquisition, and the Process of Making Travel Mode Choices." *European Journal of Social Psychology* 27, no. 5 (1997): 539–60. https://doi.org/10.1002/(SICI)1099 -0992(199709/10)27:5<539::AID-EJSP831>3.0.CO;2-A.

Verplanken, Bas, Ian Walker, Adrian Davis, and Michaela Jurasek. "Context Change and Travel Mode Choice: Combining the Habit Discontinuity and Self-Activation Hypotheses." *Journal of Environmental Psychology* 28, no. 2 (2008): 121–27. https://doi.org/10.1016/j.jenvp.2007.10.005.

Vishwanath, Arun. "Examining the Distinct Antecedents of E-mail Habits and Its Influence on the Outcomes of a Phishing Attack." *Journal of Computer-Mediated Communication* 20, no. 5 (2015): 570–84. https://doi.org/10.1111/jcc4.12126.

———. "Habitual Facebook Use and Its Impact on Getting Deceived on Social Media." *Journal of Computer-Mediated Communication* 20, no. 1 (2014): 83–98. https://doi.org/10.1111/jcc4.12100.

Volkswagen. "The Fun Theory 1—Piano Staircase Initiative." October 26, 2009. YouTube video, 1:47. https://www.youtube.com/watch?v=SByymar3bds.

———. "The Fun Theory 2—An Initiative of Volkswagen: The World's Deepest Bin." October 26, 2009. YouTube Video, 1:26. https://www.youtube.com/watch ?v=qRgWttqFKu8.

Wang, Xia, Yingying Ouyang, Jun Liu, Minmin Zhu, Gang Zhao, Wei Bao, and Frank B. Hu. "Fruit and Vegetable Consumption and Mortality from All Causes, Cardiovascular Disease, and Cancer: Systematic Review and Dose-Response Meta-Analysis of Prospective Cohort Studies." *BMJ* 349 (2014): g4490. https://doi.org/10.1136/bmj.g4490.

Wann, Daniel L., Frederick G. Grieve, Ryan K. Zapalac, Christian End, Jason R. Lanter, Dale G. Pease, Brandy Fellows, Kelly Oliver, and Allison Wallace. "Examining the Superstitions of Sport Fans: Types of Superstitions, Perceptions of Impact, and Relationship with Team Identification." *Athletic Insight* 5, no. 1 (2013): 21–44. Retrieved from http://libproxy.usc.edu/login?url=https://search .proquest.com/docview/1623315047?accountid=14749.

Wansink, Brian, and Collin R. Payne. "Eating Behavior and Obesity at Chinese Buffets." *Obesity* 16, no. 8 (2008): 1957–60. https://doi.org/10.1038/oby.2008.286.

Warren, Molly, Stacy Beck, and Jack Rayburn. *The State of Obesity: Better Policies for a Healthier America 2018*. Washington, DC: Trust for America's Health, 2018.

Wegner, Daniel M. "Ironic Processes of Mental Control." *Psychological Review* 101, no. 1 (1994): 34–52. https://doi.org/10.1037//0033-295x.101.1.34.

Wegner, Daniel M., David J. Schneider, Samuel R. Carter, and Teri L. White. "Paradoxical Effects of Thought Suppression." *Journal of Personality and Social Psychology* 53, no. 1 (1987): 5–14.

Whitehead, Alfred N. *An Introduction to Mathematics*. New York: Henry Holt, 1911.

Wiedemann, Amelie U., Benjamin Gardner, Nina Knoll, and Silke Burkert. "Intrinsic Rewards, Fruit and Vegetable Consumption, and Habit Strength: A Three-Wave Study Testing the Associative-Cybernetic Model." *Applied Psychology: Health and Well-Being* 6, no. 1 (2014): 119–34. https://doi.org/10.1111/aphw .12020.

Wing, Rena R., and Suzanne Phelan. "Long-Term Weight Loss Maintenance." *American Journal of Clinical Nutrition* 82, no. 1 (2005): 222S–25S. https://doi.org/10.1093/ajcn/82.1.222S.

Wise, Roy A. "Dopamine and Reward: The Anhedonia Hypothesis 30 Years On." *Neurotoxicity Research* 14, nos. 2–3 (2008): 169–83. https://doi.org/10.1007/bf03033808.

Wixted, John T., Laura Mickes, Steven E. Clark, Scott D. Gronlund, and Henry L. Roediger III. "Initial Eyewitness Confidence Reliably Predicts Eyewitness Identification Accuracy." *American Psychologist* 70, no. 6 (2015): 515–26. https://doi.org/10.1037/a0039510.

Wood, Wendy, and David T. Neal. "Healthy Through Habit: Interventions for Initiating and Maintaining Health Behavior Change." *Behavioral Science and Policy* 2, no. 1 (2016): 71–83. https://doi.org/10.1353/bsp.2016.0008.

Wood, Wendy, Jeffrey M. Quinn, and Deborah A. Kashy. "Habits in Everyday Life: Thought, Emotion, and Action." *Journal of Personality and Social Psychology* 83, no. 6 (2002): 1281–97. https://doi.org/10.1037//0022-3514.83.6.1281.

Wood, Wendy, Leona Tam, and Melissa Guerrero Witt. "Changing Circumstances, Disrupting Habits." *Journal of Personality and Social Psychology* 88, no. 6 (2005): 918–33. https://doi.org/10.1037/0022-3514.88.6.918.

Yin, Henry H., and Barbara J. Knowlton. "The Role of the Basal Ganglia in Habit Formation." *Nature Reviews Neuroscience* 7, no. 6 (2006): 464–76. https://doi.org/10.1038/nrn1919.

Young, Scott, and Vincenzo Ciummo. "Managing Risk in a Package Redesign: What Can We Learn from Tropicana?" *Brand Packaging* (2009): 18–21. https://www.highbeam.com/doc/1G1-208131373.html.

Zajonc, Robert B. "Attitudinal Effects of Mere Exposure." *Journal of Personality and Social Psychology* 9, no. 2 (1968): 1–27. https://doi.org/10.1037/h0025848.

Zlatevska, Natalina, Chris Dubelaar, and Stephen S. Holden. "Sizing Up the Effect of Portion Size on Consumption: A Meta-Analytic Review." *Journal of Marketing* 78, no. 3 (2014): 140–54. https://doi.org/10.1509/jm.12.0303.

Zlatevska, Natalina, Nico Neumann, and Chris Dubelaar. "Mandatory Calorie Disclosure: A Comprehensive Analysis of Its Effect on Consumers and Retailers." *Journal of Retailing* 94, no. 1 (2018): 89–101. https://doi.org/10.1016/j.jretai.2017.09.007.

Acknowledgments

I have been studying people's habits for almost thirty years and have published over a hundred articles in scientific journals. This research was so exciting, for a long time I was too engrossed to consider writing a popular book about it.

But whenever I walked into a bookstore, it was clear that someone needed to. In books written for a general audience, groundbreaking insights about the science of habits were often missing or, worse still, misconstrued. Popular books and blogs were decades behind the rapidly developing research. And each new book seemed to get further and further away from the reality we were seeing in the lab.

So I finally wrote a proposal, guided by my marvelous agent, Richard Pine at Inkwell. He helped to turn my stumbling initial descriptions into a proposal that was impressive enough to attract the support of Colin Dickerman at Farrar, Straus and Giroux. With Colin's brilliant editing and Richard's wise counsel, a book emerged; this book just wouldn't have happened if not for the guidance and constant advice of these two smart people. I also thank the talented William Callahan for making anything I sent to him delightfully more interesting.

My hundred-page proposal took a year to write, and at that point I thought I *had* to be almost done—but no. I had two more years of work and many, many discarded drafts before the manuscript finally took shape (it seems that you have to throw away a book to write a book). Through this time, I was fortunate to be funded by the University of Southern California and by the INSEAD–Sorbonne University Behavioral Lab. With the support of professor Pierre Chandon and the marketing group at INSEAD, I was awarded a position as INSEAD–Sorbonne University Distinguished Visiting Chair in Behavioral Sciences. I completed the second and third book revisions during my stay in Paris. My time at INSEAD was a great opportunity to interact with and learn from my French colleagues. The wine and cheese weren't bad, either.

The best books grow with input from many people. Most of all, I appreciate the scientists who did the remarkable research I included in my book. In addition, it was inspiring to get advice from Angela Duckworth, Jamie Pennebaker, Jonah Berger, Sam Gosling, Bob Cialdini, Tim Wilson, and Adam Grant—all award-winning scientists and writers. I continue to read their books with wonder.

As this book developed, my dear friend and longtime collaborator David Neal provided thoughtful commentary and supportive feedback ("Hey, let's meet for a glass of wine"). Generous colleagues who commented on parts of the book include Barbara Knowlton, David Neal, David Melnikoff, John Monterosso, and Bas Ver-planken. My graduate students were a continual source of inspiration, along with the talented Kristen Lee, who handled all of the references.

Families don't sign on to writing a book, and most probably wouldn't if they had a choice. Having been given no choice, my family was stalwart in its support. My dad, also a professor, loved to give advice on . . . well, pretty much everything—but specifically on writing a book. I wish he were here to see it finished, and, of course, to tease me mercilessly about any parts that didn't meet his standards. My amazing sons, Dylan and Garrett Stagner, were never too tired of hearing about the book to send me encouragement along with links to habit-related blogs and podcasts (although I admit that I still haven't listened to the end of that two-hour one). And, despite their initial discomfort at being included in this book, they ultimately re-lented and let me mention each of them exactly once.

Most of all, I thank my beloved husband, Steve Ortmann, who is the most generous person I know. He was a full partner in this undertaking, as he has been in every aspect of our lives together. This time, he had to be cheerleader, editor (who was required to love everything he read), sounding board, writing guide, and, yes, world traveler willing to quit his job and spend eight months with me in Paris. If you are wondering what I did to deserve such support, well, I wonder, too (but I'm not going to question it). Mon amour, tu est la cerise sur mon gâteau.

Index

farming, 52, 226
farm stands, 93–94
fats, 51, 55
fear, 31, 211; of cancer, 54
fiber, 51, 55
finances, *see* money
firefighting, 48–49, 50
fire safety, 140–41
fishing, 215
5 A Day for Better Health program,
    52–57
flossing, 141
flow, 215
fluency, 206; conceptual, 206;
    perceptual, 206
follow-through, 111–12
food, 7, 14–15, 17, 22, 43, 61, 124,
    136, 226–28; branded, 207;
    breakfast, 51–52, 55, 130–31, 182,
    204; cafeteria, 56, 92, 93; calories,
    148–49, 226–27; cancer and, 52, 53,
    54, 63; context and, 92–94, 95, 98;
    cooking, 21, 46, 54, 55; family meals,
    100–102, 106–107, 120, 157–58;
    farm stands, 93–94; fast food, 15,
    56–57, 208, 227; 5 A Day for Better
    Health program, 52–57; fruits and
    vegetables, 52–57, 77–79, 84, 92,
    93–94, 119–20, 144, 168, 203–204;
    habits, 15, 25, 30, 51–57, 77–79,
    119–20, 127–28, 203–206;
    hyperstimulating, 15; industry,
    15, 226; junk food, 15, 77–79;
    knowledge and, 51–57; marshmallow
    experiment, 66–69; *mise en place*,
    145–47; packaging, 168–69;
    poisoning, 126; popcorn habits, 92,
    127–28; portion size, 54–56, 226–27;
    proximity, 92–94, 168; restaurant
    buffet, 154; rewards and, 119–22,
    126; rituals, 214; self-control and,

66–69, 71, 73–74, 77–79; shopping,
    30, 93–94, 139, 168–69; swapping,
    143–44; video games, 155–57; at
    work, 100; *see also specific foods*
Food and Drug Administration (FDA),
    15
football, 48, 49–50, 210
force field principles, 90–91, 95, 106,
    154–55, 220, 231
formation, habit, 81–158; consistency
    and, 130–44; context and, 83–98;
    control and, 145–58; repetition and,
    99–114; reward and, 115–29
401(k) plan, 158
free will, 96
Freud, Sigmund, 22
friction, 90–98, 112, 154, 219, 231; cell
    phone and, 235, 236; context and,
    90–98, 147–58; control and, 145–58;
    drug use and, 198; kitchen, 145–47;
    useful, 151–52
Fritos, 144
frugality, 128
Fruits & Veggies—More Matters,
    53, 54
fruits, 52–57, 77–79, 84, 92, 93–94,
    168, 203
"Fun Theory," 119
furniture, 164

gambling, 123
Game of Thrones, 128–29
gamification, 125
Gates, Bill, 76
Geertz, Clifford, 209, 210
genetics, 190
Germany, 71
Gladwell, Malcolm, 113; 10,000-hour
    rule, 113
goals, 12, 18, 37, 38, 40, 42, 80, 89,

Illustration Credits

page 37: Decline in usage of the word "habit": Google Books Ngram Viewer, books.google.com/ngrams

page 54: More Matters logo: MoreMatters.org, © Produce for Better Health Foundation

page 58: The basal ganglia and related structures: Wikimedia Commons

page 64: Pig, duck, and goat: Pixabay

page 78: First carrot game: spiral from Pixabay; M&M's from Unsplash; carrots from Maxpixel

page 79: Second carrot game: spiral by Ernesto Kenji Salvador; M&M's from Unsplash; carrots from Maxpixel

page 136: René Magritte, *Les valeurs personnelles* (*Personal Values*), 1952: © 2018 C. Herscovici / Artists Rights Society (ARS), New York

page 146: *Mise en place*: Marcelo Trad / Shutterstock

page 156: Sushi game: composite illustration; individual images from Pixabay

page 186: Account verification email: original image

page 206: Four cars: all images from Pixabay

page 225: Map of liquor laws by county: Wikimedia Commons

page 227: The New (Ab)normal infographic: Centers for Disease Control and Prevention; use does not imply an endorsement by CDC of any product, service, or enterprise and the views expressed in the book do not necessarily represent those of CDC or HHS

page 229: Smart meter: antb / Shutterstock

A Note About the Author

Wendy Wood, Ph.D., is Provost Professor of Psychology and Business at the University of Southern California. She has written for *The Washington Post* and the *Los Angeles Times,* and her work has been featured in *The New York Times,* the *Chicago Tribune, Time* magazine, and *USA Today,* and on NPR. She lectures widely, and she recently launched the website www.goodhabitsbadhabits.org to convey scientific insight on habit to the general public.